Chromatin

Chromatin

Structure and Function

A. WOLFFE

National Institutes of Health,
Bethesda, Maryland, USA

ACADEMIC PRESS
Harcourt Brace Jovanovich, Publishers

London   San Diego   New York
Boston   Sydney   Tokyo   Toronto

This book is printed on acid-free paper

ACADEMIC PRESS LIMITED
24–28 Oval Road
London NW1 7DX

United States Edition published by
ACADEMIC PRESS INC.
San Diego, CA 92101

**A catalogue record for this book is available
from the British Library**

ISBN 0–12–761910–0
0–12–761911–9 (pbk.)

Typeset by Fakenham Photosetting Limited,
Fakenham, Norfolk
Printed in Great Britain by the University Press,
Cambridge

Contents

Preface   ix

1   **Introduction**   1

   1.1 Development of research into chromatin structure and function   2

2   **Chromatin structure**   4

   2.1 DNA and histones   4
     2.1.1   DNA structure   5
     2.1.2   The histones   11
   2.2   The nucleosome   14
     2.2.1   The nucleosome hypothesis   14
     2.2.2   The organization of DNA and histones in the nucleosome   16
     2.2.3   The structure of DNA in a nucleosome   20
     2.2.4   The position of the core histones in the nucleosome   23
     2.2.5   DNA sequence-directed positioning of nucleosomes   28
   2.3   The organization of nucleosomes into the chromatin fiber   31
     2.3.1   Histone H1 and the compaction of nucleosomal arrays   31
     2.3.2   The chromatin fiber   33
   2.4   Chromosomal architecture   39
     2.4.1   The radial loop and helical folding models of chromosome structure   39
     2.4.2   The nuclear scaffold, the centromere, telomeres, protein components and their function   43
     2.4.3   Lampbrush and polytene chromosomes   48
   2.5   Modulation of chromosomal structure   51
     2.5.1   Histone variants   51
     2.5.2   Post-translational modification of core histones   54

2.5.3   Linker histone phosphorylation   57
2.5.4   Remodeling of chromatin during
        spermatogenesis   61
2.5.5   Heterochromatin   63
2.5.6   Other structural non-histone proteins in the
        chromosome   65

3 Chromatin and nuclear assembly   68

3.1   Interactions between nuclear structure and
      cytoplasm in the living cell   68
      3.1.1   Nuclear transplantation   69
      3.1.2   Heterokaryons   71
3.2   Chromatin assembly on exogenous DNA *in vivo*   73
      3.2.1   Chromatin assembly in *Xenopus* eggs and oocytes   73
      3.2.2   Chromatin assembly on DNA introduced into
              somatic cells   78
      3.2.3   Yeast minichromosomes   82
3.3   Chromatin assembly on replicating endogenous
      chromosomal DNA *in vivo*   85
3.4   Chromatin assembly *in vitro*   88
      3.4.1   Purified systems   88
      3.4.2   Nucleosome assembly in extracts of *Xenopus* eggs
              and oocytes   90
      3.4.3   Nucleosome assembly in mammalian cell extracts   95
3.5   Nuclear assembly *in vitro*   96

4 How do nuclear processes occur in chromatin?   100

4.1   Overview of nuclear processes   100
      4.1.1   The problem of specificity   101
      4.1.2   Action at a distance   105
      4.1.3   The basal transcriptional machinery   106
      4.1.4   Stable transcription complexes   112
      4.1.5   Regulation of gene activity   116
      4.1.6   Sequence-specific DNA binding proteins   120
      4.1.7   Problems for nuclear processes in chromatin 122
4.2   Interaction of *trans*-acting factors with chromatin   123
      4.2.1   Non-specific interactions   123
      4.2.2   Specific *trans*-acting factors and non-specific
              chromatin   126
      4.2.3   Specific *trans*-acting factors and specific
              chromatin   131
      4.2.4   *Trans*-acting factors, DNaseI sensitivity, DNaseI-
              hypersensitive sites and chromosomal
              architecture   140

4.2.5 *Trans*-acting factors and the local organization of chromatin structure 147

4.3 Processive enzyme complexes and chromatin structure 150

4.3.1 Replication and the access of transcription factors to DNA 151

4.3.2 The fate of nucleosomes and transcription complexes during replication 155

4.3.3 Chromatin structure and DNA repair 159

4.3.4 Transcription and chromatin integrity *in vivo* 160

4.3.5 Transcription and chromatin integrity *in vitro* 164

5 Future prospects 167

References 170

Index 205

Preface

Research on chromatin structure and function is rapidly expanding. Technical advances allow us to follow the events regulating gene expression in the eukaryotic nucleus in molecular detail. Within the chromosome, alterations in the organization and accessibility of key regulatory DNA sequences can be documented and interpreted. This book is intended to introduce scientists to this exciting field, in the expectation that many more contributions will be required before we understand completely how the nucleus of a eukaryotic cell functions.

The book has five chapters. Chapter 1 gives a brief overview of the issues discussed and an historical account of their development. In Chapter 2 the structure of chromatin and chromosomes is described as far as it is known. Concepts concerning chromatin structure are already very well-developed, indeed many of the biophysical techniques and paradigms for studying protein–nucleic acid interactions were pioneered using the basic unit of chromatin, the nucleosome, as a model. In contrast, large-scale chromosomal architecture is much less well defined, as is the influence of modifications of structural proteins on chromatin and chromosome organization. How these changes contribute to the various requirements for correct chromosomal function is a recurring theme.

A complete understanding of the eukaryotic nucleus requires not only that we know how to take it apart, but also that we can assemble it from the various component macromolecules. Chapter 3 describes the approaches, results and interpretations of experiments designed to accomplish this task. The biological constraints of assembling a chromosome rapidly are discussed with reference to its final form and properties.

Form and function are intimately related, once a complete understanding of a process is achieved, it is impossible to separate one from the other. Chapter 4 describes the multitude of approaches taken towards resolving how DNA can be folded into a chromosome and yet still remain accessible to the regulatory proteins, and allow processive enzymes to move along the length of the DNA molecule. It is in this field of research that much of the current progress on the interrelationship of chromatin structure and function is taking place. The final chapter offers a perspective on where prospects for future development might lie.

I would like to thank participants in the NIH chromatin group for sharing their ideas and results, especially Drs Trevor Archer, David Clark and Sharon Roth. I am indebted to Drs Randall Morse, Geneviève Almouzni, Jeffrey Hayes and my editor, Dr Susan King, for their comments on the text. Appreciation and thanks are given to Ms Thuy Vo and Mr William Mapes for preparing the manuscript and figures. Finally I thank my wife Elizabeth for her patience and support during the preparation of this book.

Alan Wolffe

CHAPTER ONE

Introduction

Our knowledge of how the hereditary information within eukaryotic chromosomes is organized and used by a cell has increased enormously through the application of recombinant DNA methodologies. Technical advances now allow individual DNA sequences to be isolated and their association with proteins within the cell nucleus to be determined. Experimental progress has led the biologist to explore long-standing questions concerning how a particular cell acquires and maintains its individual identity. Developmental biologists have used the techniques of molecular biology and genetics to investigate how an egg differentiates into different cell types. These questions have led scientists to the realization that growth, development and differentiation proceed through regulated changes in the form and composition of specific complexes of protein and DNA within the nucleus. Understanding how these complexes are assembled and function has become a central theme in modern biology.

Many of the techniques used to probe protein–DNA interactions were developed by researchers interested in the basic structural matrix of chromosomes – chromatin. This complex of DNA, histones and non-histone proteins has been exposed to a multitude of biochemical, biophysical, molecular biological and genetic manipulations. The structure of chromatin is by now well-understood, but how it is folded and compacted into a chromosome is not. Knowledge of how chromatin is constructed preceded the development of methods capable of exploring function. The purification and cloning of non-histone proteins required to perform the complex events involved in DNA transcription, replication, recombination and repair is the focus of a continuing and intense research effort. Investigators now have the tools to try to integrate our experience with chromatin structure and assembly with the structural proteins and enzymes required for the maintenance, expression, and duplication of the chromosome.

1.1 DEVELOPMENT OF RESEARCH INTO CHROMATIN STRUCTURE AND FUNCTION

Towards the end of the nineteenth century numerous investigators formulated the theory that chromosomes determined inherited characteristics (see Voeller, 1968). These studies were almost entirely based on cytological observations with the light microscope. Although chromosomes are clearly only present in the nucleus, the influence of components of the cytoplasm on inherited characteristics was examined by forcing embryonic nuclei into regions of the cytoplasm in which they would not normally be found (Wilson, 1925). These experiments and others led Morgan (1934) to propose the theory that differentiation depended on variation in the activity of genes in different cell types. The genes were clearly in the chromosomes, but their biochemical composition remained completely unknown.

The last quarter of the nineteenth century also saw the recognition of RNA (first identified as yeast nucleic acid), DNA (thymus nucleic acid) and the discovery of histones. Albrecht Kossel isolated nuclei from the erythrocytes of geese and examined the basic proteins in his preparations, which he named the histones (reviewed by Kossel, 1928). The apparent biochemical simplicity of DNA and the obvious complexity of protein in chromosomes led investigators mistakenly to regard the latter component as the major constituent of the elusive genes (Stedman and Stedman, 1947). Only the gradual acceptance of experiments on the capacity of DNA alone to change the genetic characteristics of the cell (Avery *et al.*, 1944) led to the recognition of nucleic acid as the key structural component of a gene.

The elucidation of the double helical structure of DNA with its immediate implications for self-duplication, opened up the new approaches of molecular biology to clarifying the nature of genes (Watson and Crick, 1953). Although the double helix was now recognized as containing the requisite information to specify a genetic function, how this information was controlled was not understood.

The apparent heterogeneity of the histones due to proteolysis and the various modifications of these proteins suggested that they might be important in regulating genes. Eventually methodological improvements for isolating and resolving the different histones demonstrated that they were highly conserved in eukaryotes and that only a few basic types existed (Fitzsimmons and Wolstenholme, 1976). This lack of variety implied that histones themselves were unlikely to be the primary regulators of gene activity.

During the 1970s the organization of the fundamental complex of DNA and histones, which came to be called the nucleosome, was largely solved. Recombinant DNA methodologies facilitated the isolation and cloning of defined DNA sequences, and DNA sequencing enabled the *cis*-acting elements potentially controlling gene expression to be defined (Brown, 1981). Elucidation of the organization of regulatory DNA was immediately followed during the 1980s by the vigorous search for the non-histone proteins – the *trans*-acting factors that might interact and function at these regulatory elements. Many of these proteins have now been characterized (Johnson and McKnight, 1989). This endeavor has led to the current focus of molecular biology and genetics: the attempt to understand how *trans*-acting factors and the enzyme complexes involved in DNA replication, transcription, recombination and repair function *in vivo*. The continual improvement in experimental techniques has led to the realization that important regulatory elements in DNA are organized into specific structures including both the histone proteins and *trans*-acting factors. Much current research concerns this interrelationship, the structure of these complexes and their function in the utilization of the hereditary information within the chromosomes.

CHAPTER TWO

Chromatin Structure

Chromosomes represent the largest and most visible physical structures involved in the transfer of genetic information. Surprisingly, our understanding of chromosome organization is most complete for the smallest and most fundamental structural units. These units are the nucleosomes which contain both DNA and histones. The long folded arrays of nucleosomes along the axis of a chromosome comprise the vast majority of chromatin. In this section we will discuss the structural features of DNA and histones, how they assemble into nucleosomes and how nucleosomes fold into chromatin fibers. Finally I will describe what we know about the organization of the chromatin fiber into a chromosome and how this can be modified in various ways.

2.1 DNA AND HISTONES

The most striking property of a chromosome is the length of each molecule of DNA incorporated and folded into it. The human genome of 3×10^9 bp would extend over a meter if unravelled, however this is compacted into a nucleus of only 10^{-5} m in diameter. It is an astonishing feat of engineering to organize the long linear DNA molecule within ordered structures that can reversibly fold and unfold within the chromosome. Not surprisingly, many aspects of chromosome structure reflect the impediments and constraints imposed by having to bend and distort DNA.

2.1.1 DNA structure

DNA has an elegant and simple structure around which the chromo-some is assembled. The DNA molecule exists as a long unbranched double helix consisting of two antiparallel polynucleotide chains. DNA always contains an equivalent amount of the deoxyribonucleo-tide containing the base adenine (A) to that with the base thymine (T), and likewise of the deoxyribonucleotide containing the base gua-nine (G) to that with the base cytosine (C) (Fig. 2.1). Each base is

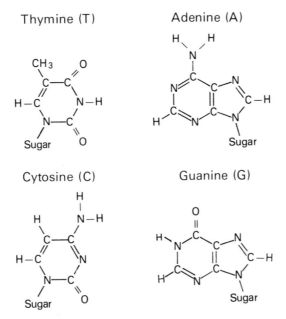

Figure 2.1.   The four bases found in DNA.

linked to the pentose sugar ring (2-deoxyribose) and a phosphate group. The 5′ position of one pentose ring is connected to the 3′ position of the next pentose ring via the phosphate group (a 5′–3′ linkage) to create the polynucleotide chain (Fig. 2.2). The two antipar-allel polynucleotide chains are attached to each other by hydrogen bonding between the bases. G is always base paired to C, and A is always base paired to T. In addition to the stability imparted by hydrogen bonding, hydrophobic base stacking interactions occur along the middle of the double helix (Fig. 2.3) (see Lewin, 1990 or Alberts *et al.*, 1990 for details).

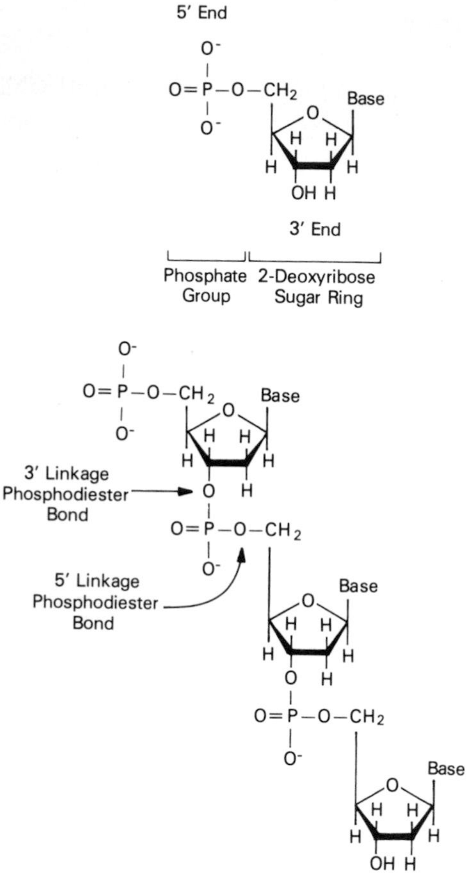

Figure 2.2.   A nucleotide and a polynucleotide chain.

Physical studies using X-ray diffraction indicate that under conditions of physiological ionic strength, DNA is a regular helix, making a complete turn every 3.4 nm with a diameter of 2 nm. This particular DNA structure is known as B-DNA and has approximately 10.5 bp/turn of the helix. This means that every base pair is rotated approximately 34° around the axis of the helix relative to the next base pair. This results in a twisting of the two polynucleotide strands around each other. A double helix is formed that has a minor groove (approximately 1.2 nm across) and a major groove (approximately 2.2 nm across). The geometry of the major and minor grooves of DNA will be seen later to be crucial in determining the interaction of proteins with the DNA backbone. The double helix is right handed (Fig. 2.4).

Beyond this basic description, DNA structure is exceedingly plastic.

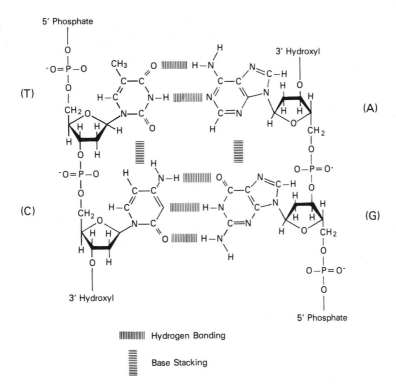

Figure 2.3.   The interactions stabilizing the two antiparallel polynucleotide chains in DNA.

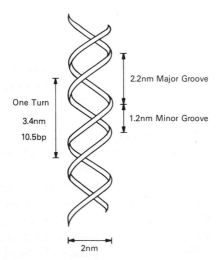

Figure 2.4.   The dimensions of DNA.

Crystallization of various oligonucleotides indicates that a variety of DNA sequences will yield recognizable B-form DNA structures (Privé *et al.*, 1991; Yanagi *et al.*, 1991). More severe alterations in the conditions under which DNA is examined do, however, generate distinct conformations. Dehydrating the fiber will cause the double helix to take up a structure known as A-DNA (11 bp/turn); or placing DNA with a defined sequence of alternating G and C bases in solutions of high ionic strength will lead to the formation of a left-handed helix known as Z-DNA (12 bp/turn). The existence of either of these extreme structures in the eukaryotic nucleus under normal physiological conditions is controversial. However their formation indicates the gross morphological changes that DNA can be forced to undergo (Drew *et al.*, 1988).

How do we know what structure populations of DNA molecules have in solution? Two experimental methodologies have been commonly used. The first employs DNA cleavage reagents and a flat crystal surface (Rhodes and Klug, 1980). When DNA is absorbed from solution onto a flat calcium phosphate surface and cut with DNaseI, the enzyme cuts DNA most readily where it is exposed away from the surface. The average spacing between the sites of cleavage gives the approximate number of base pairs per turn of DNA (Fig. 2.5). This is determined by the electrophoresis of denatured molecules through a polyacrylamide gel. A better reagent for this purpose is the hydroxyl radical. Hydroxyl radicals are generated by the Fenton reaction in which an Fe(II) EDTA complex reduces hydrogen peroxide to a hydroxide anion and a hydroxyl radical. The radical is about the size of a water molecule and has little sequence specificity in cleaving DNA. This it does by breaking the pentose sugar rings of individual deoxyribonucleotides. In contrast, DNaseI is a large enzyme which has considerable sequence preferences. In both instances, the number of base pairs per turn of a large population of different DNA sequences bound to a crystal surface is found to be 10.5 (Tullius and Dombroski, 1985). This result is consistent with DNA having a B-form configuration as determined by X-ray studies.

The second method to examine DNA structure in solution reaches the similar conclusion that DNA has a B-form conformation at physiological ionic strength; however, a completely different strategy is used. It is generally found that a population of closed circular DNA molecules, identical in length and sequence contains different numbers of superhelical turns. The formation of superhelical turns can be simply described as follows: a single superhelical turn is introduced into a closed circular DNA molecule if the molecule is broken, one end of the molecule is then fixed, the other is rotated once and

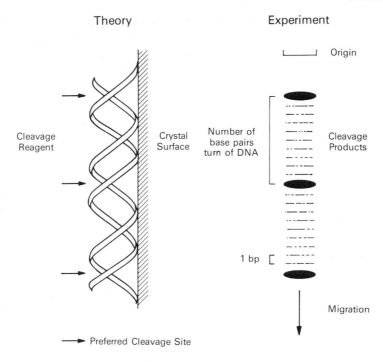

Theory   Experiment

Origin

Cleavage
Reagent

Crystal
Surface

Number of
base pairs
turn of DNA

Cleavage
Products

1 bp

Migration

Preferred Cleavage Site

Figure 2.5.   Determining the helical periodicity of DNA in
'solution' through binding to a flat crystal surface and cleavage
with an enzyme or a chemical reagent.

In theory the most exposed region of the double helix will be cut preferentially, experimentally this is reflected in a larger population of DNA fragments cut at this site after resolution on a polyacrylamide gel (darker
bands). The distance between darker bands in base pairs is the helical
periodicity (number of base pairs per turn) of DNA.

the two ends rejoined. Supercoils can be positive or negative depending on which way the free DNA end is rotated. Closed circular molecules of the same length and sequence with different numbers of
superhelical turns are known as topoisomers. Each population of
small closed circular DNA molecules that differ in length by a few
base pairs will exist as a distribution of topoisomers. These can be
resolved by electrophoresis through an agarose gel matrix. A molecule which has a length corresponding to an integral number of
helical turns will exist predominantly as a single topoisomer whereas
a molecule which deviates from this by half a helical turn will be
equally likely to exist with the superhelical turn in a positive or negative sense. The number of DNA molecules with a particular mobility
in the agarose gel will be reduced by half since the molecules exist as
an equal mixture of topoisomers. Examining the relationship between

DNA length and the distribution of topoisomers allows the number of base pairs per turn of DNA to be calculated. The result of 10.5 bp/turn is close to that derived from crystal binding studies (Horowitz and Wang, 1984). Finally, theoretical calculations of the most stable configuration of DNA, that actually preceded much of the experimental work, suggested a value of 10.6 bp/turn (Levitt, 1978). The range of values around 10.5 bp/turn obtained both experimentally and theoretically, provide a sound basis for considering alterations in this structure based on DNA sequence content and histone–DNA interaction.

Aside from the dramatic changes in DNA structure seen on formation of A- or Z-DNA, local variations in DNA sequence can significantly influence DNA conformation and properties of the helix. Our most extensive knowledge of the local changes in B-form DNA structure due to sequence content come from studying AT-rich DNAs. For example, oligo(dA).oligo(dT) tracts are found experimentally using both spectroscopic techniques and DNA cleavage reagents such as the hydroxyl radical, to be straight and rigid with a constant narrow minor groove width (Nelson *et al.*, 1987; Hayes *et al.*, 1991a). This is believed to be a consequence of maximizing the hydrophobic base stacking interactions between adjacent A.T base pairs in the DNA helix (Fig. 2.3). This stabilization process requires the bases to be more twisted relative to each other than would normally be found in typical B-form DNA. Chains of these base pairs have the correct geometry to allow at least two water molecules per base pair to become highly ordered along the DNA backbone. This creates a 'spine of hydration' which contributes to the rigidity of oligo(dA).oligo(dT) tracts (Berman, 1991). Changes in sequence that affect these structural features lead to widening of the minor groove, for example, a G.C base pair will disrupt the straight path and rigidity of an oligo(dA).oligo(dT) tract. In contrast to oligo(dA).oligo(dT), oligo[d(A–T)] tracts are conformationally flexible. This flexibility is a consequence of not being able to achieve efficient hydrophobic base stacking interactions between consecutive T.A and A.T base pairs without severely distorting the DNA helix (Travers and Klug, 1987; Travers, 1989). Finally, short oligo(dA).oligo(dT) tracts (4–6 bp in length) that are phased with a periodicity similar to that of the DNA helix itself, will cause the molecule to be curved. This is due to a narrowing of the minor groove every turn of DNA caused by the phased oligo(dA).oligo(dT) tract (Koo *et al.*, 1986). Periodicities that are greater or smaller than 10–11 bp will cause the normally straight DNA to take on a 'corkscrew-like' path. In spite of this wide variation in 'B-form' DNA structure, all of these DNA sequences can be assembled into chromatin (Section 2.2.5).

Summary

Under most physiologically relevant conditions DNA is a stable B-form double helix, with 10.5 bp/helical turn, a major and a minor groove. Local variations in sequence content can direct DNA to have intrinsic rigidity, flexibility or curvature.

2.1.2 The histones

The primary proteins whose properties mediate the folding of DNA into chromatin are the histones. Aside from the compaction of DNA, the histone proteins undertake protein–protein interactions between themselves and other distinct chromosomal proteins. These interactions lead to several constraints on the properties of histones contributing to maintaining their high degree of evolutionary conservation. Not all eukaryotic cells have histones, for example, dinoflagellates package their DNA with small basic proteins completely unlike histones (Vernet *et al.*, 1990); and in mammalian species the majority of DNA in spermatozoa is compacted through interaction with basic proteins known as protamines (Section 2.5.4). Certain eukaryotes are deficient in particular histones, for instance, yeast cells are apparently deficient in the type of histone that binds to DNA between nucleosomes (Section 2.3.1).

Each nucleosome contains a core of histones around which DNA is wrapped. This core contains two molecules of each of four different histone proteins: H2A, H2B, H3 and H4. These are known as the core histones. Since histones can be removed from DNA by high salt concentrations, the major interactions between DNA and the core histones appear to be electrostatic in nature. Histones H2A and H2B dissociate first as the salt concentration is raised followed by histones H3 and H4 (see Section 2.2.2). Studies of this type coupled to chemical cross-linking, demonstrated that histones H2A and H2B form a stable dimer (H2A/H2B), whereas histones H3 and H4 form a stable tetramer $(H3/H4)_2$ in the absence of DNA (Kornberg, 1974; Kornberg and Thomas, 1974). Many histone proteins have been purified and their amino acid sequences determined. Subsequently, histone genes have been cloned and a very complete picture of core histone sequence properties established.

All core histones are remarkably conserved in length and amino acid sequence through evolution. Histones H3 and H4 are the most highly conserved; for example, calf and pea histone H4 differ at only two sites in 102 residues (DeLange *et al.*, 1969a, b). Histones H3 and H4 have a central role in both the nucleosome and consequently many chromosomal processes (Sections 2.2.2 and 2.2.4); these functional and structural requirements presumably contribute to their remarkable sequence conservation. Histones H2A and H2B are slightly less conserved. All of the core histones are small basic proteins (11,000–16,000 Da molecular weight) containing relatively large amounts of lysine and arginine (more than 20% of the amino acids). Histones H2A and H2B contain more lysine (14 out of 129, and 20 of of 125 amino acids respectively in calf), and histones H3 and H4 contain more arginine (18 out of 135, and 14 out of 102 amino acids respectively in calf) (van Holde, 1989). All four histones contain a globular domain through which histone–histone and histone–DNA interactions occur, and charged tails which contain the bulk of the lysine and arginine residues (Fig. 2.6). Like the globular regions of the

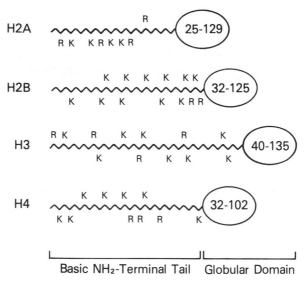

Figure 2.6.   The organization of calf thymus histones.
The N-terminal histone tails are shown as zig-zag lines with lysine (K) and arginine (R) residues indicated. Globular regions are indicated by the ellipsoids.

histones, which might be expected to be conserved due to their central structural role in the nucleosome, the amino acid sequence of the charged tails is also conserved. These charged tails are the sites of many post-translational modifications of the histone proteins (Section 2.5.2). Core histone variants in which the primary amino acid sequence is slightly changed because of expression of different alleles of a histone gene and post-translational modifications of the histone tails have important consequences for chromatin structure and function in many contexts, but especially during development (Sections 2.5.2 and 2.5.4).

Most eukaryotic cells contain a fifth histone called the linker histone, of which the most common is called histone H1 In addition, many studies have examined the properties of a specialized linker histone from chicken erythrocytes known as histone H5. Both histone H1 and histone H5 are highly basic, being particularly rich in lysine and are slightly larger than core histones (> 20,000 molecular weight) (Fig. 2.7). They have a central globular domain and highly charged

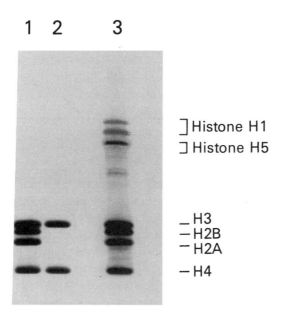

Figure 2.7.   The histones are shown resolved on a denaturing polyacrylamide gel, separated by virtue of their size.
Core histones (H3, H2B, H2A and H4) and linker histones (H1, H5) are indicated. Histones were prepared from chicken erythrocytes.

tails at both the amino (N-) and carboxyl (C-) terminus. Linker histones are not found within the histone core. Instead, the globular domain interacts with DNA at one side of the nucleosome where it begins and finishes wrapping around the core (Section 2.5.2). In addition, the linker histone tails interact with the DNA between nucleosomes. It is possible that these tail regions fold up into α-helices that lie in the major groove of DNA. Linker histones are the least tightly bound of all histones to DNA, and are readily dissociated by solutions of moderate ionic strength ($>$ 0.35 M NaCl). They are the most variable in amino acid sequence and hence structure, and are also extensively post-translationally modified both during the cell cycle and during development (Section 2.5.3). These structural modification have important consequences for the functional properties of the chromatin fiber.

Summary
Two types of histones exist, the highly conserved core histones and the variant linker histones. Both types have a globular domain and highly charged basic tails. These tail regions are the site of post-translational covalent modifications.

2.2 THE NUCLEOSOME

The nucleosome is the fundamental repeating unit of chromatin. Many of the techniques used to examine protein–nucleic acid interactions that are in common use today, were pioneered on the nucleosome. Outlining how the current model of the nucleosome has been developed will introduce the use of nucleases and chemical probes both of DNA structure and protein–DNA interaction (DNA footprinting reagents), non-denaturing gels to study large complexes of protein and DNA in their native state (mobility shift assays), together with various applications of spectroscopic analysis and other biophysical techniques.

2.2.1 The nucleosome hypothesis

The first clear insights into the nucleosomal organization of chromatin came from nuclease experiments (both intended and accidental) in which the DNA in chromatin was found to degrade to a series of discrete fragment sizes separated by multiples of 180–200 bp (Williamson, 1970; Hewish and Burgoyne, 1973). Each step in fragment size is now known to represent the DNA associated with a single nucleosome. Extensive nuclease digestion allowed each DNA fragment to be isolated as a complex with protein (Sahasrabuddhe and van Holde, 1974). These particles were found by sedimentation analysis in the analytical ultracentrifuge to have a mass of around 200,000 Da (176,000 was measured) of which the protein content was close to 110,000 Da (105,000 was measured). This we now know to correspond to the octamer of core histones in a nucleosome (two molecules each of histones H2A/H2B/H3 and H4) plus approximately 146 bp of DNA.

Electron microscopic analysis of chromatin provided further evidence for a structure consisting of discrete complexes of protein and nucleic acid arrayed along the DNA backbone. The pictures of 'beads on a string' were compelling evidence for a repeating particulate structure for chromatin (Olins and Olins, 1974). Each particle along the DNA backbone was approximately 10 nm in diameter, similar to that of the particles isolated by extensive nuclease digestion. Chemical cross-linking experiments led to the realization that the core histones existed in a precise stoichiometry (Kornberg and Thomas, 1974).

These observations, over a few years in the early 1970s, led to the proposal by Kornberg of the nucleosome model (Kornberg, 1974). This hypothesis suggested that each particle consisted of DNA and histones. DNA was wrapped around an octamer of the core histones, each octamer consisting of a tetramer of histones H3 and H4 ((H3/H4)$_2$) and two dimers of H2A/H2B (H2A/H2B). Initially it was thought possible that only a small fraction of chromatin in the nucleus might be organized in this way. However, the structural significance of the nucleosomal organization of chromatin was made clear by micrococcal nuclease digestion of whole nuclei. These studies revealed that over 80% of DNA in the nucleus was incorporated into nucleosomes (Noll, 1974a). Thus, the general relevance of the nucleosome for the folding of DNA in the eukaryotic nucleus was firmly established. Subsequent studies have considerably refined our understanding of its organization.

Summary
Studies involving nuclease digestion, analytical centrifugation, electron microscopy and chemical cross-linking led to the proposal that a fundamental repeating unit of chromatin existed, consisting of a precise stoichiometry of histones and DNA. This particulate structure became known as the nucleosome. The vast majority of DNA in the cell nucleus is organized into nucleosomes.

2.2.2 The organization of DNA and histones in the nucleosome

Like many scientific fields, the study of chromatin has developed a particular nomenclature: a *nucleosome* consists of one repeating length of DNA in the nucleus, generally determined by very slight micrococcal nuclease cleavage (see below), plus an octamer of core histones and a single molecule of linker histone; a *nucleosome core particle* consists of the octamer of core histones and the length of DNA that resists digestion even after extensive micrococcal nuclease cleavage, this being the 146 bp of DNA that have the strongest contacts with the histone core. The DNA that is in between nucleosome core particles, and that is lost when a nucleosome is trimmed to a core particle, is called linker DNA (Fig. 2.8).

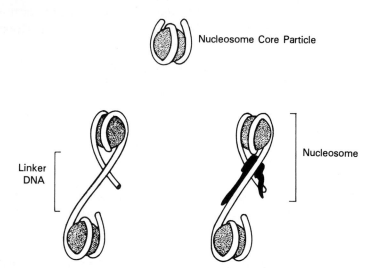

Figure 2.8.   The organization of a nucleosome core particle, a nucleosome and the position of linker DNA are shown.
Core histones form the shaded spheres, DNA is the connecting tube and the linker histone is the solid shape.

Most nucleosomes have been isolated for study by micrococcal nuclease cleavage. Micrococcal nuclease cleaves chromatin in the most accessible DNA (Fig. 2.9). First the linker DNA between nucleosomes is cut, then the nuclease digests the rest of the linker before it cuts DNA directly across both strands within the nucleosome itself. The nucleosome, representing the first product of very slight micrococcal nuclease digestion will therefore be progressively trimmed to a

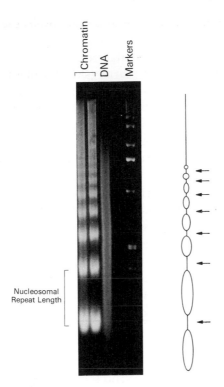

Figure 2.9.   Micrococcal nuclease cleavage of chromatin.
Chromatin and naked DNA are shown after treatment with micrococcal nuclease, removal of protein and resolution on an agarose gel. The nucleosome ladder of bands is clearly seen in the chromatin samples. Cleavage of naked DNA generates a wide distribution of DNA fragments visible as a smear. The visible bands in the chromatin samples are separated from each other by a single nucleosome repeat length of DNA.

nucleosome core particle as digestion with micrococcal nuclease proceeds. The nucleosome core particle itself represents only a kinetic intermediate in the digestion of DNA. Eventually micrococcal nuclease can degrade the DNA in this residual structure and the core particle will fall apart. Separation of nucleosomes from nucleosome core particles demonstrated that the initial digestion of the linker DNA led to the loss of linker histone from the nucleosome (Noll and Kornberg, 1977).

The next advance in understanding nucleosomal structure came from the use of non-denaturing gel electrophoresis to examine large complexes of DNA and proteins in their native state. The mobility of free DNA is retarded through association with protein, producing a mobility shift. Following micrococcal nuclease digestion, crude nucleosomal fractions were first resolved on sucrose gradients. The smallest (mononucleosome) fractions were then electrophoresed through a polyacrylamide gel matrix in a mobility shift assay, and two complexes were resolved. The large (slower migrating) complex was found to contain histone H1, whereas the smaller complex did not (Varshavsky *et al.*, 1976). Subsequently, careful gradation in the extent of nuclease digestion allowed Simpson (1978) to isolate a discrete particle, called the chromatosome, consisting of an octamer of histones, one molecule of the linker histone H1 and about 160 bp of DNA.

The influence of histone H1 on nucleosome integrity was examined using biophysical techniques. Spectroscopic experiments on chromatosomes examined DNA conformational transitions in these particles dependent on increasing temperature in comparison to particles depleted of histone H1. Thermal denaturation of DNA requires the progressive disruption of both hydrogen bonds and base stacking interactions between the base pairs in DNA, and eventually the separation of the two strands of the DNA helix ('melting'). Histone H1 was found to significantly stabilize DNA within the nucleosomal core particle (Simpson, 1978). These experiments led to the conclusion that histone H1 interacted not only with the linker DNA but also with the DNA wrapped around the histone octamer. These interactions are important both for the folding of nucleosomal arrays into the chromatin fiber and for the interaction of *trans*-acting factors with DNA (Sections 2.3.2 and 4.2.2).

Having defined the site of interaction of the linker histone H1 with DNA in the nucleosome, we will now discuss the organization of the nucleosome core particle itself. Early studies of the organization of DNA in chromatin used nucleosomal core particles obtained by extensive micrococcal nuclease digestion of nuclei. These particles presumably contain representatives of every DNA sequence present in nucleosomes within the nucleus (> 80%, see Section 2.2.1). Using the enzyme DNaseI Noll found that single nicks were made in the DNA of the core particles with a periodicity of 10 bp. This periodicity of cleavage reflects the wrapping of DNA around the histone core. Compare these results with those from the nuclease cleavage of DNA when bound to a crystal surface (see Section 2.1.1). This wrapping implies that the minor groove of the double helix, which is recognized and cut by DNaseI (Drew, 1984), is exposed only once per turn of the helix (Fig. 2.10). Noll found that the entire length of the core particle is exposed to nicking by DNaseI, which means that all 146 bp of DNA must lie on the surface of the histone core (Noll, 1974b).

Noll's cleavage data for the organization of DNA in the nucleosome

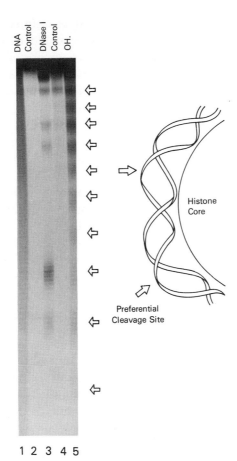

1 2 3 4 5

Figure 2.10. DNaseI and hydroxyl radical cleavage of DNA in the nucleosome.

DNA is made radioactive at one end, by the introduction of a phosphate group containing ^{32}P. Hydroxyl radical cleavage of naked DNA is shown (DNA). Control lanes are nucleosome core particles that do not have a cleavage reagent added to them. In the DNaseI and OH • lanes the indicated cleavage reagent is added. The DNA fragments are denatured to single strands and are then resolved on a denaturing polyacrylamide gel on the basis of size. Where DNA is exposed away from the histone core it is cut as indicated in the line drawing.

were indirect. The most substantial evidence for DNA being wrapped around a histone core prior to crystallization, came from neutron scattering in solution. These experiments relied on the observation that the intensity of neutron scattering by a particular macromolecule depends on the intensity of scattering by the solvent in which the macromolecule is bathed. By adjusting the scattering of the solvent through altering the ratio of heavy water (D_2O) to normal water (H_2O), certain macromolecules in solution could be made invisible. Thus, by looking at the scattering of DNA or protein in the nucleosome in aqueous solutions, it was found that DNA had a larger radius of gyration than the protein component. Therefore DNA was wrapped around the histones (Pardon *et al.*, 1975).

The observation that DNA lies on the surface of the nucleosome was greatly extended in detail by the crystallization of nucleosome core particles by Klug and colleagues (Finch *et al.*, 1977; Richmond *et al.*, 1984). Analysis of the crystals to 7Å resolution revealed the nucleosome to have the disc-like shape predicted from electron microscopy, sedimentation analysis and neutron scattering. The disc was 11 nm in diameter and 5.5 nm in height. DNA was wrapped in 1.75 turns of a left-handed double helix around the histone core. However, the bending of the DNA around the histone core was not uniform. There were very sharp bends one and four helical turns to either side away from the center of the nucleosomal DNA. The nucleosome appears almost symmetric, hence the center of nucleosomal DNA also represents the dyad axis of the nucleosome. Although it is still not clear whether part of the distortion of DNA is due to crystallization artefacts, the overall bending of DNA and potential perturbations of the path of the double helix from a uniform curve have important implications for the phenomenon of nucleosome positioning (Section 2.2.5). The organization of the individual core histones within the nucleosomal core particle is not yet known from the crystal structure; more indirect methods have had to be used to investigate this problem (Section 2.2.4).

Summary
Linker histones, such as histone H1, bind to both the linker DNA and the DNA associated with the histone octamer. DNA (146 bp) is wrapped in 1.75 left-handed superhelical turns around an octamer of the core histones.

2.2.3 The structure of DNA in a nucleosome

The structure of DNA in a nucleosome can be refined from the 7Å crystal structure using chemical probes. Independent evidence for deformation of the DNA double helix came from using highly reactive singlet oxygen. This reagent was found to preferentially react with DNA about 1–1.5 turns either side of the center of the nucleosomal

DNA in solution (Hogan *et al.*, 1987). Other evidence for a deformation of DNA at this position came from hydroxyl radical cleavage of DNA in nucleosomal core particles. Single base pair resolution analysis of hydroxyl radical cleavage frequencies in nucleosomal DNA reveals that the central three turns of DNA in the nucleosome have a different number of base pairs per turn (10.7) than the remainder of the structure (10.0) (Fig. 2.11). When these two different structures

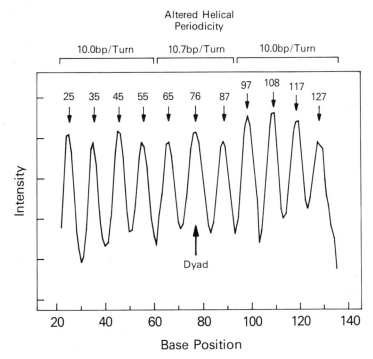

Figure 2.11. DNA has two different structures in the nucleosome. The three turns of DNA around the dyad axis have a different helical periodicity to those in the remainder of the nucleosome.
A plot of hydroxyl radical cleavage frequency at each nucleotide of the DNA within nucleosome core particles is shown. Approximate positions of maximum cleavage frequency are indicated. Numbers indicate the absolute length of the single-stranded DNA fragments.

juxtapose, DNA is likely to be distorted. These hydroxyl radical cleavage results also imply that the structure of DNA changes on association with a histone core from that in solution (10.5 bp/turn) (Hayes *et al.*, 1990, 1991b). This has important consequences for the association of sequence specific DNA binding proteins (Sections 4.2.2 and 4.2.3).

As 146 bp of DNA are present in the nucleosome core particle, a simple calculation based on the hydroxyl radical cleavage results predicts that DNA wrapped around the histone core will have an average number of base pairs per helical turn of 10.2. This is in substantial agreement with several other independent measurements of this important value. Synthetic DNA curves designed to have different helical periodicities (see Section 2.1.1), reveal a preference of the histone core for a 10.2 bp/turn structure overall (Shrader and Crothers, 1990). DNaseI cleavage analysis of DNA in the nucleosome revealed an average cleavage frequency of 10.4 bp/turn; however, the periodicity of DNaseI cleavage sites at the two ends of DNA in nucleosomes was close to 10.0 bp, while the spacing of cleavage sites towards the center was closer to 10.6 bp/turn (Lutter, 1978). Interpretation of these results as directly reflecting the helical periodicity of DNA in a nucleosome core particle is not possible. This is due to the large enzyme, DNaseI being sterically hindered from having equivalent access to DNA all around the histone core by the turn of DNA adjacent to the one it is cutting (Klug and Lutter, 1981). However, the use of other methodologies such as examination of the frequency of thymidine dimer formation (photofootprinting) in nucleosomal DNA, reveals modulation with a periodicity of 10.0 bp/turn on either side of a central region, which has a periodicity of about 10.5 bp/turn (Gale and Smerdon, 1988) (Section 4.3.3). The most gruelling approach to this problem has required sequence analysis of DNA within 177 nucleosome core particles. Quantitation of the sequence data reveals periodic modulations in the frequency with which certain runs of three base pairs are found. The phases of these periodic modulations are offset by 2–3 base pairs across the central 2–3 turns of the nucleosome (Satchwell *et al.*, 1986). All of these experiments strongly suggest a change in helical periodicity for the central region of nucleosomal DNA.

What is precisely responsible for the distortion of the central portion of DNA in the nucleosome is unknown, although this is the site of strong contacts with both histone H3 molecules in the histone core (Section 2.2.4). The distortion of DNA from 10.5 to 10.0 bp/turn in nucleosome core particle DNA away from the central region is better understood. Theoretical calculations by Levitt predicted a 10.6 bp/turn structure for DNA free in solution and a 10.0 bp/turn structure for DNA constrained into an 80 bp circle as found in the nucleosome (Levitt, 1978).

The change in DNA structure on incorporation into a nucleosome from 10.5 to an average of 10.2 bp/turn partially explains a long-standing problem that has come to be known as the 'linking number paradox' (Klug and Lutter, 1981). This follows from the fact that the nucleosome has at least 1.75 superhelical turns of DNA around the histone core, yet the measured number of superhelical turns introduced into a relaxed DNA molecule (in the presence of topoisomerases) is 1. This discrepancy is partially explained by the change in helical periodicity of DNA upon incorporation into a nucleosome, resulting in a decrease in the average number of base pairs per turn. This overwinding of DNA can account for the disappearance of 0.4 superhelical turns, but does not completely explain the 'paradox' (Hayes *et al.*, 1990). A more contorted path than the wrapping of DNA around a simple cylinder could account for the small differences remaining (White *et al.*, 1988).

Summary

DNA in the nucleosome is severely distorted into two circles of approximately 80 bp in length. The helical periodicity of this DNA changes from that in solution. Two regions of 10.0 bp/turn flank the three central turns of nucleosomal DNA which have a periodicity of 10.7 bp/turn (average for the nucleosomal core particle equals 10.2 bp/turn).

2.2.4 The position of the core histones in the nucleosome

Where are the individual core histones along the DNA backbone in the nucleosome core particle? Early studies indicated that the histone H3/H4 tetramer, $(H3/H4)_2$ could organize DNA into nucleosome-like particles. Hydroxyl cleavage radical patterns of DNA in such particles demonstrates that the tetramer organizes the central 120 bp of DNA of a nucleosome identically to that of the nucleosomal core particle (Hayes *et al.*, 1991b; Fig. 2.12). This type of footprinting analysis with subnucleosomal particles agrees well with chemical cross-linking experiments which physically map histone–DNA contacts along nucleosomal DNA. Most of the chemistry for this type of approach has been developed by Mirzabekov (Mirzabekov *et al.*, 1977, 1982).

Chemical cross-linking of histones to DNA is generally performed after the 5' end of the DNA within nucleosome core particles has been labeled with ^{32}P so that the position of histone–DNA contacts relative to the end of the molecule (as a reference point) can be determined. This is followed by reaction with dimethylsulfate which methylates purine bases. The methylated product is then depurinated to an aldehyde. A Schiff base is formed between the modified DNA backbone

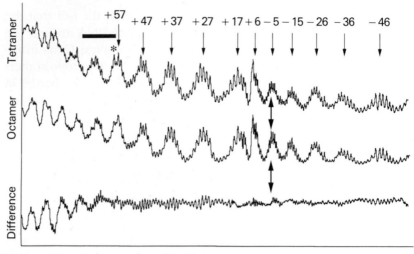

Figure 2.12. A comparison of DNA interactions with a complete octamer of core histones to those with the histone tetramer (H3/H4)$_2$.

A nucleosome structure including the 5S RNA gene is used. Densitometric analyses of hydroxyl radical cleavage patterns of the complexes of the tetramer and octamer are shown, together with a difference plot. The position of the peaks in hydroxyl radical cleavage with respect to position +1 of the 5S RNA gene (numbers) and the dyad axis (bold vertical arrows) are shown. Asterisk indicates the position of the first reproducible difference between the two patterns. Horizontal bar indicates the region where the octamer pattern diverges from that of the tetramer (see Hayes *et al.*, 1991b for details).

and available lysine amino acids in the histones, which can be further stabilized by reduction with sodium borohydride (Fig. 2.13). Very low levels of reaction are allowed so that a wide distribution of DNA–histone cross-links are generated. The DNA–histone complex is denatured and the histone–DNA adducts are resolved from each other by electrophoresis. The mobility of the DNA molecule is reduced by the cross-linked protein. Resolution in a second dimension after removal of histones using a protease, allows sizing of DNA pieces. The relative mobility of protein-bound and free DNA permits the organization of the histones relative to the end of DNA in the nucleosomal core particle to be determined.

Several important conclusions emerge from this analysis. A major result was the realization that the histones bind to nucleosomal DNA in a symmetrical linear array. Histone H3 is found to have weak interactions with DNA where it enters and leaves the core particle. We will see later that the amino acids involved in these contacts may be modified, with important consequences (Section 2.5.2). Moving

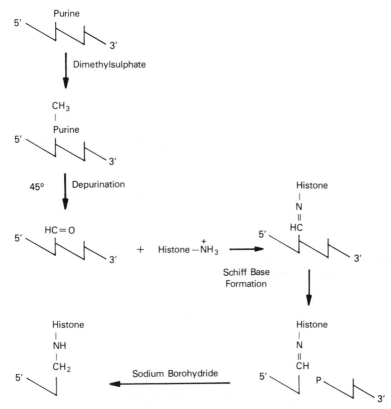

Figure 2.13. Chemistry of cross-linking histones to DNA.

away from the ends of nucleosomal DNA, histones H2A/H2B bind at the periphery and histones H3/H4 towards the center (Fig. 2.14). Particularly strong protein–DNA contacts occur where histone H3 appears to distort the central turns of DNA in the nucleosome, perhaps accounting for the change in helical periodicity in this region (Section 2.2.3). These observations are completely consistent with all of the biophysical results (see Section 2.2.2).

Knowing the location of the histones along the DNA backbone does not tell us whether the globular domain or the highly charged tails of the proteins are contacting the double helix. Several lines of experimental evidence suggest it is not the histone tails, but the globular domains of the histones that actually organize DNA in the nucleosome. It is possible to remove the positively charged histone tails from nucleosome core particles without altering the accessibility of DNA to DNaseI or the hydroxyl radical (Fig. 2.15, Hayes *et al.*, 1991b). DNA conformation is not changed by removal of the tails. The

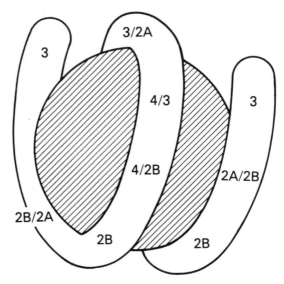

Figure 2.14.   Positions of core histone contacts with DNA as it wraps around the histone core.
Histones are indicated by their numbers, where two histones contact the same region of DNA, both are shown.

stability of the trypsinized nucleosome core particles to physical perturbations, such as increased temperature or high salt concentrations, reveals that the proteolysed particle undergoes no major changes in stability in comparison to intact particles. This suggests that the histone tails have little or no role in maintaining the integrity of the nucleosome (Ausio *et al.*, 1989).

Although at first sight it might be thought that the histone tails would associate with the major groove of DNA wrapped around the nucleosome core particle, experimental evidence has indicated otherwise. Dimethylsulfate has been used to probe the accessibility of bases within the major groove of DNA in the nucleosome. Results indicate that there is almost uniform reactivity with this small molecule, suggesting that the tails do not bind in the major groove (McGhee and Felsenfeld, 1979). Moreover, DNA that has glucose groups within the major groove still forms apparently normal nucleosomes, even though access by proteins to base pairs would be severely hindered (McGhee and Felsenfeld, 1982). These experiments lead to the conclusion that the histone tails have little or no role in organizing DNA in the nucleosome core particle. Instead they have a distinct and equally important role in protein–protein interactions outside of the nucleosome (Section 2.5.2).

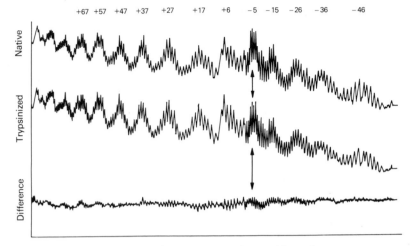

+67 +57 +47 +37 +27 +17 +6 −5 −15 −26 −36 −46

Figure 2.15. A comparison of DNA interactions with an intact octamer of core histones to those with an octamer from which the histone tails have been removed with trypsin.

A nucleosome structure including the 5S RNA gene is used. Densitometric analyses of hydroxyl radical cleavage patterns of the complexes with the intact and trypsinized octamer are shown, together with a difference plot. The position of the peaks in hydroxyl radical cleavage with respect to position +1 of the 5S RNA gene (numbers) and the dyad axis (bold vertical arrow) are shown (see Hayes *et al.*, 1991b for details).

Focusing on the globular domains of the core histones, considerable evidence supports electrostatic interactions of the phosphodiester backbone of DNA with arginine residues present in the histone octamer as being most important for organizing DNA in the nucleosome. The accessibility of arginine and lysine residues to low molecular weight chemicals (2,4,6-trinitrobenzoic acid and 2,3-butanediol respectively) in various subnucleosomal particles reveals that only 14 arginine residues are required to maintain the integrity of the globular region of the nucleosome core particle. This major role for arginine may be due to the capacity of this residue to form both hydrogen-bonding and electrostatic interactions with phosphate residues along the DNA backbone (Ichimura *et al.*, 1982).

The apparent simplicity of the histone–DNA interactions maintaining the integrity of the nucleosome core particle has some interesting consequences. It has long been known that nucleosome core particles can be dissociated into their DNA and histone components by elevating ionic strength. This suggests that the primary interactions responsible for the stability of the particle are electrostatic. Histones H2A/

H2B dissociate first, at salt concentrations above 0.8 M NaCl, whereas histones H3/H4 only dissociate from DNA when the concentration rises above 1.2 M NaCl (Ohlenbusch *et al.*, 1967). An important aspect of this process is that it is reversible. This means that on dilution of these high ionic strength salt solutions, core histones will reassociate with any available DNA to reassemble nucleosomal structures. In a strict nomenclature, these structures are not nucleosomes, because they have no histone H1, nor are they nucleosome core particles because they contain more than 145 bp of DNA. I have chosen to call them *nucleosome structures*. Much of our current understanding of nucleosomal organization derives from analysis of these synthetic nucleosome structures (Section 2.1.5).

Summary
Biophysical analysis and chemical cross-linking reveals that the tetramer of histones H3/H4 has the central role in organizing the nucleosome. Histones H2A/H2B interact to either side of the tetramer and consequently with DNA towards the ends of the molecule as it wraps around the histone core. The globular regions of the core histones are responsible for organizing DNA around the histone core, primarily through electrostatic interactions between arginine residues and the phosphodiester backbone (Fig. 2.16).

2.2.5 DNA sequence-directed positioning of nucleosomes

Most current research into the organization of DNA in the nucleosome relies on methodologies to reassemble nucleosomal structures onto defined DNA templates. Early experiments of this type demonstrated that histone cores can not interact efficiently with double stranded RNA or RNA–DNA heteroduplexes to form nucleosomal structures. Extensive homopolymeric stretches (> 60 bp) of rigid oligo(dA).oligo(dT) or of the left-handed Z-DNA duplex are also not favored for incorporation into nucleosomes (Prunell, 1982; Garner and Felsenfeld, 1987) (see Section 2.1.1). All of these results reflect the energetic cost of deforming the unusual nucleic acid structures (A-form, rigid DNA or Z-form) into the bent 10.0 bp/turn structure of nucleosomal DNA. These represent extreme circumstances, since most naturally occurring DNA sequences – including the AT-rich sequences discussed earlier (Section 2.1.1) – are incorporated into a nucleosome at little energetic cost (Hayes *et al.*, 1991a). However, the

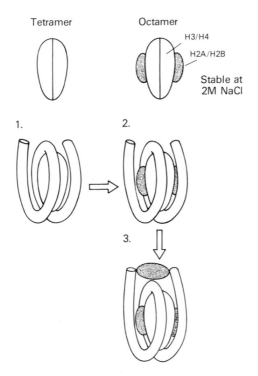

Figure 2.16. A simple view of the nucleosome.
The (H3/H4)$_2$ tetramer is shown as an open ellipsoid, H2A/H2B dimers (stippled ellipsoids) interact to either side of the (H3/H4)$_2$–DNA complex (1), thereby incorporating more DNA into the particle (2). When two turns of DNA are wrapped around the core histones, histone H1 (horizontal ellipsoid) can associate (3).

local influences of DNA rigidity and curvature will affect the precise translational and rotational position the double helix adopts with respect to the histone core. This phenomenon is known as nucleosome positioning and has important biological consequences (Section 4.2.3).

The translational position of a nucleosome refers to where the histone core starts and finishes being associated with DNA. The rotational position refers to which face of the double helix is in contact with, or exposed away from, the histone core. Nucleases and chemical probes have been very useful in determining both of these parameters. Considerable evidence exists in support of sequence-directed positioning of nucleosomes on DNA. For example, chicken and frog core histones *in vitro* and yeast histones *in vivo* all recognize the same structural features of a 5S ribosomal RNA gene which direct nucleosome position (Fig. 2.17; Simpson, 1991). Mutagenesis experiments in

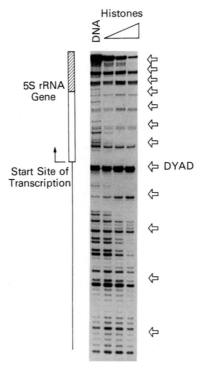

Figure 2.17.   A positioned nucleosome includes the site of transcription initiation for the *Xenopus* 5S RNA gene.

The gene is shown as an open box, key regulatory elements inside the gene as a hatched box, and the start of transcription as a hooked arrow. In this experiment DNA is radiolabeled at one end, it is then cleaved with DNaseI either as naked DNA (lane 1) or as a nucleosome (lanes 2–4). DNA is then denatured, single strands separated with respect to size on an acrylamide gel and autoradiographed. Major cleavage sites and the dyad axis of the nucleosome are indicated.

which DNA sequences in and around the 5S RNA gene were perturbed indicated that a region comprising 20–30 bp to either side of the center of nucleosomal DNA contained the elements necessary for positioning (FitzGerald and Simpson, 1985). This region contains the sharp bends and structural discontinuities in DNA structure observed within the nucleosome (Section 2.2.3).

   The nucleosome positioning sequence defined for the 5S ribosomal RNA gene has a DNA structure with a periodic modulation in minor groove width. If such modulations occur every helical turn of DNA, as they do for the 5S nucleosome, positioning element, this directs the DNA molecule to have an intrinsic curvature (Hayes *et al.*, 1990). It is energetically favorable to incorporate a DNA sequence that is already curved into a nucleosome, as DNA has to be curved around the histone core anyway (Shrader and Crothers, 1989). Narrowed minor grooves will face into the nucleosome and wide minor grooves will face out (Fig. 2.10). In the region about 10–15 bp from the mid-point of the nucleosome core particle DNA, the minor groove varies from 0.7 nm on the inner face of the helix to 1.3 nm on the outside (Morse and Simpson, 1988). A consequence of this is that synthetic DNA curves have proven very useful for directing nucleosome position both translationally and rotationally (Wolffe and Drew, 1989). However, not all nucleosome positioning is sequence directed as non-histone proteins can also influence the association of histones with DNA (Section 4.2.5).

Summary
Due to the structural variations in DNA within the nucleosome, rigid or curved DNA sequences will influence the position of the histone octamer along the DNA backbone (translational nucleosome positioning). This will lead to the double helix having a particular face orientated towards the histone core (rotational nucleosome positioning).

2.3 THE ORGANIZATION OF NUCLEOSOMES INTO THE CHROMATIN FIBER

Although it is relatively easy to visualize the array of nucleosomes along the DNA molecule, this represents only the first level of compaction of DNA in the nucleus. Our understanding of the compaction of DNA into higher order structures than nucleosomal arrays is much weaker than our knowledge of the nucleosome. This probably reflects not only unavoidable heterogeneity in the naturally occurring structures, but also the difficulty in isolating and studying large macromolecular complexes by conventional techniques.

2.3.1 Histone H1 and the compaction of nucleosomal arrays

A key molecule in directing the formation of higher order structure in a nucleosomal array is the linker histone H1. However, histone H1 is not essential for chromatin folding. The folding of histone H1-deficient chromatin provides us with important insights into how and why nucleosomal arrays are compacted. Several studies using 'natural' chromatin depleted of histone H1 and synthetic chromatin in which nucleosome structures are positioned at varying distances apart, have clearly demonstrated that nucleosomal arrays can be compacted simply by varying the concentration of mono- and divalent cations in solution (Clark and Kimura, 1990; Hansen *et al.*, 1989). Under appropriate conditions it is possible to compact chromatin to the level observed *in vivo* in the presence of histone H1. These results indicate that the limiting factor in chromatin compaction is the degree of shielding of charge along the phosphodiester backbone of DNA. An important question is whether this shielding involves all of the DNA in the nucleosome or just the linker DNA between nucleosomes. Comparison of the salt-induced folding of chromatin in the presence or absence of histone H1, suggests that H1 only has to interact with the linker DNA to facilitate the compaction of nucleosome arrays (Clark and Kimura, 1990). Although this is a simple and important result that probably underlies most of folding of nucleosomes into higher order structures, there is considerable specificity in the interaction of histone H1 with DNA in the nucleosome and important consequences concerning other aspects of the folding of nucleosomal arrays.

The first experiments to examine the role of histone H1 in chromatin structure concerned its disappearance from the nucleosome as linker DNA is digested away by micrococcal nuclease, and the isolation under controlled digestion conditions of the chromatosome containing the histone octamer, about 160 bp of DNA and a single molecule of histone H1 (Section 2.2.2). The symmetry of the nucleosome core particle suggests that this single molecule of histone H1 would interact with an extra 10 bp either side of the central 146 bp of the nucleosomal core particle. However, this has yet to be rigorously established. Histone H1 is an asymmetric molecule, which on interaction with a nucleosomal core particle will create an asymmetric nucleosome. Experiments with proteolytic fragments of histone H1 suggest that the globular domain of histone H1 binds where DNA enters and exits the nucleosome and across the few central turns of DNA in the structure (Figs. 2.8 and 2.16; Allan *et al.*, 1980, 1981, 1986; Staynov and Crane-Robinson, 1988). This interaction at the periphery of the particle seals two turns of DNA around the histone octamer, leading to further stabilization of the interaction of DNA with the octamer (Simpson, 1978).

Extended fibers of chromatin containing histone H1 have a zigzag structure in the electron microscope consistent with the points of entry and exit of DNA from individual nucleosomes being close together (Worcel *et al.*, 1978). In histone H1-depleted chromatin, the fibers are more extended, which would be expected if the entry and exit points of DNA wrapping around the octamer were far apart (Thoma *et al.*, 1979). The globular domain of histone H1 also exhibits considerable selectivity for binding to supercoiled rather than linear DNA (Singer and Singer, 1976). Supercoiled DNA is constrained such that the double helix crosses over itself more frequently than seen with relaxed circular or linear molecules. This selectivity is consistent with the globular domain interacting with the ends of DNA wrapped around the nucleosome core particle and with this domain having at least two DNA binding surfaces. In addition to this interaction with DNA organized by the core histones, the long C-terminal tail domains of histone H1 fold into α-helices that associate with the major groove of the linker DNA between nucleosomes (Clark *et al.*, 1988). In contrast to the preference of the histone H1 globular domain for supercoiled DNA, the histone H1 tails bind to both linear and supercoiled DNA without preference. Thus, histone H1 binds to DNA and core histone-associated DNA in an organized manner.

It is possible to specifically cross-link histone H1 to the core histones (histone H2A) using reagents that will create covalent bonds only between peptide chains that are in intimate contact (zero-length) (Boulikas *et al.*, 1980). This linkage between histones H1 and H2A implies that direct protein–protein contacts exist; however, the significance of this communication for the stability to the nucleosome is unknown. It is likely that either these contacts or others with the histone tails are important since a link between the post-translational modification of the core histones and histone H1 binding is well-established (see Section 2.5.2).

Summary

Although nucleosomal arrays will fold in its absence, histone H1 binding to linker DNA will facilitate the process. Histone H1 constrains DNA where it enters and leaves wrapping around the histone core. Direct protein–protein contacts with a core histone (histone H2A) also exist.

2.3.2 The chromatin fiber

Much information concerning the higher order organization of nucleosomal arrays has followed visualization of chromatin fragments prepared under various conditions in the electron microscope

(Thoma *et al.*, 1979). These observations show that at very low salt
(0.2 mM EDTA, 1 mM triethanolamine chloride) chromatin appears as
a zigzag fiber of nucleosomes. At slightly higher salt (0.2 mM EDTA,
5 mM triethanolamine chloride) a flat ribbon forms about 25 nm wide,
whereas at moderate ionic strengths (100 mM NaCl, approaching
physiological relevance) chromatin condenses to form irregular rod-
like structures with a diameter of about 30 nm (Fig. 2.18). These 30 nm

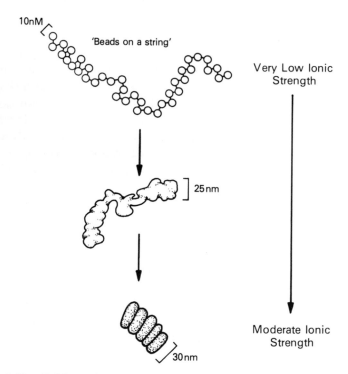

Figure 2.18.   Folding of chromatin as visualized in the electron
microscope.
As ionic strength increases chromatin changes from a 'beads-on-a-string'
configuration to a flat fiber, 30 nm in diameter.

fiber structures are similar to those that have been observed in nuclei
prepared by a variety of techniques (see Section 2.4.1), hence the
interchangeable description of the chromatin fiber as the 30 nm fiber.
Much of the chromatin in the cell nucleus of higher eukaryotes is now
believed to exist in this configuration. How the nucleosomal arrays
are precisely folded into the chromatin fiber is unresolved. Almost as
many models have been proposed as experiments have been carried

out. Every model will not be reviewed in detail, but the focus will be on the proposals for which the most extensive experimental support exists.

As described above, Thoma and colleagues used the electron microscope to examine the unfolding and folding of long chromatin fragments containing several kilobase pairs of DNA, dependent on ionic conditions. Based on their observations they proposed a model for the structure of the 30 nm fiber in which the chain of nucleosomes is compacted by winding into a simple solenoidal structure. In this model there would be approximately six nucleosomes per helical turn and the pitch of each turn would be about 11 nm. The pitch was decided on the basis of visible 11 nm striations in the rod-like 30-nm-wide structures seen in electron micrographs (Fig. 2.18). Each turn of the solenoid might therefore correspond to the nucleosomes stacking with their long axes parallel to the fiber, since the nucleosomes are individually shaped like 11-nm-wide discs (Thoma *et al.*, 1979). An implicit part of this model is that the folding of nucleosomal arrays might be promoted by interaction between the linker histones down the axis of the fiber. Histone H1 is known to bind cooperatively to naked DNA and it has been presumed that comparable interactions exist in chromatin (Clark and Thomas, 1986). These might be assisted by the asymmetry of the histone H1 molecule, imparting a directionality to the fiber axis.

Other evidence supports the folding of nucleosomal arrays being mediated through interactions between linker histone molecules down the central axis of the chromatin fiber. Experiments examining the accessibility of linker histones to antibodies and proteases are consistent with a reduced accessibility of these proteins when an array of nucleosomes is compacted (Losa *et al.*, 1984; Dimitrov *et al.*, 1987). Of course the decrease in accessibility of linker histones might occur in other ways than sequestration within the central axis of the chromatin fiber. However, chemical cross-linking studies have clearly demonstrated that significant changes in the interaction of linker histones with DNA occur during chromatin compaction. This modified interaction of linker histones with DNA in chromatin differs from that seen with naked DNA in isolation. This implies that protein–protein contacts within chromatin can influence the structure and perhaps the mode of interaction of histone H1 with DNA (Mirzabekov *et al.*, 1990). Such changes would be consistent with contacts between histone H1 molecules being generated only in the chromatin fiber.

Aside from the ribbon and rod-like conformations of chromatin fibers, other electron microscopic images reveal different structures including those of globular clusters of nucleosomes ('superbeads'). Once again it is possible to detect bead-like discontinuities in the chromatin fiber in sectioned nuclei. Under certain conditions these superbeads can be remarkably uniform, containing between 8 and 48 nucleosomes. This suggests they represent discrete structural units (Zentgraf and Franke, 1984). However, the failure to find a ubiquitous organization of chromatin into superbeads indicates that they are the consequence of partial breakdown or rearrangement of a continuous solenoidal fiber. This may, in fact, represent the true state of affairs within the nucleus, as we will discuss later; it would, in fact, be surprising if eukaryotic DNA was packaged away into structures of a crystalline order and stability.

The simple solenoid proposed by Thoma *et al.* is not the only helical model for the folding of nucleosomal arrays (Felsenfeld and McGhee, 1986). A major alternative that has received considerable attention builds on the observation that zigzag arrays of nucleosomes exist in solutions of low ionic strength and that these arrays are dependent on the presence of linker histones. In models developed by Worcel, Woodcock and colleagues, the zigzag array forms a condensed ribbon containing two parallel rows of nucleosomes, the folding once again mediated by linker histone. Coiling of this ribbon generates the 30 nm fiber (Woodcock *et al.*, 1984).

Various spectroscopic techniques have been used to test these different proposals for the compaction of nucleosomal arrays. Neutron scattering was applied with a similar philosophy to that outlined earlier for the nucleosome (Section 2.2.2), to measure the distribution of DNA in the chromatin fiber. The observed constraints on the dimensions of the fiber support solenoid models in which the nucleosomes are positioned radially, like the spokes of a wheel (Suau *et al.*, 1979). Among the other techniques used is electric dichroism, which requires chromatin fibers to be orientated in electric fields and their absorbance of polarized ultraviolet light to be measured. Since the path of DNA in the nucleosome and the electric dichroic properties of DNA are well-understood, it is possible to calculate the expected properties of any array of orientated nucleosomes. Felsenfeld, Charnay and colleagues showed that a key assumption of the solenoid models was correct, in that the nucleosomes were orientated with their long axes parallel to the fiber (McGhee *et al.*, 1980, 1983). Fiber diffraction studies also lend further support to the simple solenoid models. Widom and Klug obtained partially orientated chromatin fibers by drawing concentrated solutions of chicken erythrocyte chromatin into capillaries (Widom and Klug, 1985). Analysis of the diffraction data again suggested radial packing of nucleosomes; more importantly bands that might be predicted by two parallel rows of nucleosomes forming a solenoid (Worcel *et al.*, 1978; Woodcook *et al.*, 1984) were absent. Further consideration of this type of model led Felsenfeld and McGhee to propose a modification of the simple solenoid model, in which linker DNA is not constrained in the center of the 30 nm fiber but simply follows the path of the chain of nucleosomes itself. In this 'coiled linker' model, the linker DNA is coiled between adjacent nucleosomes (Fig. 2.19; Felsenfeld and McGhee, 1986).

Coiled Linker DNA

Figure 2.19.   The coiled linker model of the chromatin fiber. Three turns of only one side of the fiber are shown.

For any of these models the length of DNA in a nucleosomal repeat is very important. This is because the length of linker DNA constrains the type of arrangement one nucleosome can have relative to the other. In all organisms and tissues the size of DNA that is resistant to micrococcal nuclease digestion in a nucleosome core particle is remarkably constant at about 146 bp. However the distance between the first cuts by micrococcal nuclease, i.e. the nucleosomal repeat length of DNA in the nucleosomal core particle plus linker DNA, varies greatly. In *Saccharomyces cerevisiae* the nucleosomal repeat length is merely 165 bp, whereas in most mammalian cell lines it is 180–200 bp in length, and grows to almost 260 bp in sea urchin sperm (van Holde, 1989). Variation in the length of linker DNA leads to clear experimental differences in both the efficiency with which different chromatin preparations will compact and in the properties of the resultant chromatin fiber.

Langmore and colleagues have shown, using chromatin with long nucleosome repeat lengths, that the diameter of the chromatin fiber increases with DNA linker length (Athey *et al.*, 1990; Williams and Langmore, 1991). At the other extreme, Widom has demonstrated the efficient compaction of *Saccharomyces* chromatin with very short repeat lengths. Nucleosomal arrays containing linkers of less than 20 bp are perfectly able to fold into a 30 nm fiber (Lowary and Widom, 1989). More importantly, linker DNA that is very short compared to expectations of DNA persistence length (or 'stiffness'), has been shown to bend in chromatin. Histone H1 clearly increases the flexibility of linker DNA by neutralizing negative charge on the phosphodiester backbone (Section 2.3.1); however, it was unclear whether this neutralization would be sufficient for every short linker DNA to bend. Light scattering, which measures the radius of gyration of a particle, and electron microscopy, which gives a direct visualization of particle size, have been used to demonstrate compaction of dinucleosomes (containing histone H1) and dinucleosomal structures (depleted of histone H1). The properties of these defined chromatin substrates were identical to those predicted from the compaction studies of long chromatin with or without histone H1 (Yao *et al.*, 1990, 1991). This flexibility of linker DNA lends strong support to the 'coiled linker' modification of the simple solenoid model proposed by Felsenfeld and McGhee.

Although the available evidence favors the simple solenoid model for the folding of most nucleosomal arrays, it is quite possible that other ways of folding occur. Langmore and colleagues have clearly documented several exceptions to the simple 30 nm fiber (Athey *et al.*, 1990). Unlike DNA in the nucleosome, considerable flexibility and heterogeneity are likely to exist in the higher order structure of chromatin. This complexity in the folding of chromatin is increased by strong influences on the structure due to various post-translational modifications of both core and linker histones that occur during the normal cell cycle and in development (Sections 2.5.2 and 2.5.3).

Summary
Most of the available evidence supports a solenoidal folding of nucleosome arrays into the chromatin fiber (30 nm fiber). The linker DNA appears to be coiled between nucleosomes in this structure.

2.4 CHROMOSOMAL ARCHITECTURE

The chromatin fiber is not necessarily a static, stable structure (Section 2.3.2), as proteins including histones continually equilibrate in and out of these structures (Section 3.1). Properties of the fiber can also be changed by modification of its constituent histone proteins or through interaction with structural non-histone proteins that recognize the fiber and package it into a chromosome. The chromatin fiber is organized into large domains potentially separated through interaction with a nuclear scaffold or matrix. These chromosomal domains are themselves folded in an ordered manner to form the chromosome.

2.4.1 The radial loop and helical folding models of chromosome structure

Our most thorough understanding of chromosomal organization is for the most condensed and hence most visible of chromosomes, those at metaphase. Although folding of DNA into nucleosomes leads to a seven-fold compaction in length, and the subsequent folding of arrays of nucleosomes into the chromatin fiber to a further seven-fold compaction, a massive 250-fold compaction of DNA follows the organization of the chromatin fiber into a metaphase chromosome (Earnshaw, 1988). Two principle models have been proposed to account for this compaction (Fig. 2.20). The first suggests an organization of the fiber into loops that are radially arranged along the axis of the chromosome. The second suggests a helical folding of the chromatin fiber, followed by a helical folding of the resultant 250 nm fiber (Sedat and Manuelidis, 1978).

These two models have received a great deal of attention in recent years. The evidence for the organization of the chromatin fiber into loops attached to a central axis in normal cells comes from several experimental approaches. There are long-standing observations on the morphology of lampbrush chromosomes in amphibian oocytes. Here a succession of loops are seen emerging from a single chromosomal axis (Section 2.4.3). Worcel and Burgi (1972) had developed a model for the *E. coli* chromosome that predicted its organization into independent domains or loops. Worcel extended these studies to interphase chromosomes from *Drosophila* cells. Intact chromosomes were subjected to very mild digestion with DNaseI, so as to produce single-strand nicks. The sizes of the resulting chromosomal fragments were then examined in sedimentation experiments and found to

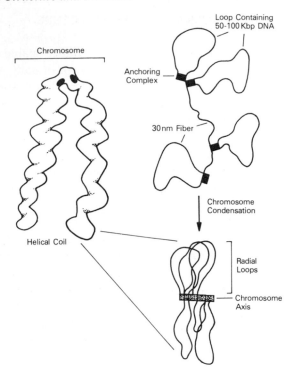

Figure 2.20.   The folding of loops of the chromatin fiber into the chromosome.
The chromatin fiber is tethered at anchoring complexes which interact with each other to form the chromosome axis.

decrease gradually until a plateau value of fragment size was reached in which the complexes contained approximately 85,000 bp of DNA complexed with protein (Benyajati and Worcel, 1976). This suggested that the chromatin fiber was organized into fairly uniform domains containing about 100 kbp of DNA. A second approach involves the electron microscopy or sedimentation of nuclei that are extracted of histones by exposure to high salt solution. It is possible to directly measure the length of DNA on the microscope grid from where it exits a residual nuclear structure (nucleoid, now known as the nuclear scaffold or matrix) to where it re-enters this structure. Estimates of loop size between 40 and 90 kbp were obtained, consistent with the biochemical measurements (Cook and Brazell, 1975; Jackson *et al.*, 1990).

The development of pulsed-field gel electrophoresis (Schwartz and Cantor, 1984) allowed a more systematic analysis of the separation of cleavage sites following mild nuclease digestion of nuclei (Filipski *et al.*, 1990). This technique allows the resolution of DNA molecules of very large size by agarose electrophoresis. Nuclei of eukaryotic cells contain an endonuclease (first responsible for the discovery of the nucleosomal repeat, see Section 2.2.1) that can be activated under controlled conditions (by the addition of exogenous Ca^{2+}/Mg^{2+} to nuclei) and which is believed to have little sequence specificity. It might be expected that cleavage of DNA by this enzyme would be inhibited by folding of DNA into the chromatin fiber, but that structural discontinuities, perhaps where the loop is attached to the chromosome axis might allow cleavage if the enzyme was activated. Activation of this enzyme followed by resolution of the resultant DNA fragments on pulse field gels would potentially allow loop sizes to be determined. A second approach to this problem makes use of the observation that topoisomerase II is a major component of the nuclear scaffold (Section 2.4.2). The enzymatic action of topoisomerase II is to introduce a double strand break into DNA which is then resealed (Table 2.1). These double strand breaks can be stabilized by

Table 2.1. Eukaryotic DNA topoisomerases

	Activity
Topoisomerase I (Topo I)	Removes superhelical turns from DNA in the absence of an energy source; this is done by introducing a transient break in one strand and allowing rotation of the broken strand about the intact DNA chain, resulting in alterations in linking number by steps of one
Topoisomerase II (Topo II)	Removes superhelical turns from DNA in an ATP-dependent reaction; this is done by breaking both strands and allowing rotation of both strands about the intact DNA chain, resulting in alterations in linking number by steps of two

the use of specific drugs (e.g. epipodophyllotoxins) that inhibit the enzyme. Preferential cleavage sites in nuclei spaced 50–300 kb apart were detected using both the endogenous nuclease or topoisomerase II in the presence of the specific inhibitors of rejoining of the double helix. Closer analysis revealed a hierarchy of digestion, where the 300 kb cleavage products appeared before those of 50 kb. It has been suggested that the 50 kb intermediate in digestion represents the first level of organization of the chromatin fiber into loops, whereas the 300 kb kinetic intermediate in digestion represents the next level of organization. Taken together these observations establish a strong case for large independent loops (50–100 kb) of the chromatin fiber representing a unit of chromosome structure.

Support for the second model proposing the folding of the chromatin fiber without specific attachments to an undisrupted chromosome axis, comes from sophisticated high-voltage and conventional transmission electron microscopy combined with extensive computer analysis by Sedat, Agard and colleagues. The 'scaffolding' hypothesis would suggest that although the arrangement of loops of the chromatin fiber about the chromosome axis might form distinct patterns, no discrete higher order organization exists above the chromatin fiber. Sedat and Agard demonstrate convincingly that such an order does in fact exist (Belmont *et al.*, 1987, 1989). Early work on large plant chromosomes during the meiotic cycle had suggested that coils of chromatin fibers existed in the chromosome (White, 1973). Under certain fixation conditions chromosomes from animal cells could also appear as a spiral or zigzag fiber (Onnuki, 1968). Sedat and Agard examined native *Drosophila* mitotic chromosomes, observing a hierarchy of higher order chromatin folding patterns for the 30 nm fiber: structures of 50, 100 and 130 nm in diameter were clearly discernible. Although a looping architecture of the 30 and 50 nm fibers could be detected under certain circumstances, the loops were not observed to be consistently orientated radially in three dimensions about any given axis, and no evidence for a central scaffolding was apparent. These results are not inconsistent with an important role for non-histone scaffolding proteins in anchoring local loops or domains of chromatin structure, but do appear to rule out a strict radial symmetry or undisrupted central axis for such loops. Most likely a diffuse organization of loops anchored to non-histone proteins actually exists *in vivo* (Fig. 2.20).

Consistent with this concept are immunofluorescent experiments in which antibodies against topoisomerase II reveal it to be distributed in separate 120–200 nm islands (Section 2.4.2). However, Sedat and Agard also found no evidence to support the helical folding of the most condensed form of chromatin, the 130 nm fiber, to form the some. This appears to exclude their earlier models (Sedat and Manuelidis, 1978). Instead they propose a model whereby mitotic chromosomes would progress out of and into interphase through the association and dissociation of distinct 130-nm-wide domains of chromatin. Although the exact structure of these large chromatin domains is even more uncertain than that of the 30 nm chromatin fiber, it is clear that non-histone proteins must be important for directing these particular aspects of chromosome organization.

Summary
Considerable evidence from both nuclease digestion biophysical, and

morphological studies supports the formation of large loops of the 30 nm chromatin fiber in the eukaryotic nucleus. However these loops do not appear to originate from a central continuous axis. Instead, high resolution light microscopy suggests that the axis can continually assemble and disassemble. Loops of the 30 nm chromatin fiber appear to be organized into even more complex structures to assemble the chromosome.

2.4.2 The nuclear scaffold, the centromere, telomeres, protein components and their function

Many studies have focused on the non-histone proteins present in the nuclear scaffold or matrix and the DNA sequences associated with them. Nuclear scaffolds (or matrices) can be prepared by several different extraction procedures using high salt or detergents which remove most of the histone and non-histone proteins. The fact that nuclear scaffolds existed at all was initially objected to on several grounds. The first objection was that an axial scaffolding of a chromosome had never been observed in normal mitotic chromosomes. This objection can be dealt with by assuming that the scaffold represents an aggregation of discrete anchoring complexes that may be connected within the chromosome by being assembled on the same DNA molecule (Section 2.4.1). The second problem came from the methodologies used for preparing nuclear scaffolds. The scaffold represents the fraction of cellular proteins that is insoluble under the extraction procedures employed. Why should a biochemist believe that these proteins necessarily had anything to do with chromosomes at all? Perhaps they were just adventitiously precipitated by the extraction procedure used. However scaffolds of apparently identical polypeptide compositions are obtained when chromosomes are extracted at high ionic strength (2 M NaCl), with the polyanions dextran sulfate and heparin at low ionic strength ($\sim$ 10 mM NaCl), or with low concentrations (5 mM) of the chaotropic agent lithium diiodosalicylate. It is very unlikely that these different treatments would precipitate the same population of chromosomal non-histone proteins (Earnshaw, 1988).

Several distinct proteins were initially identified in chromosome scaffold preparations. These include the lamins A, B and C which are intermediate filament-like proteins that constitute the nuclear lamina (Franke, 1987). This is the fibrillar protein meshwork that lines the nucleoplasmic surface of the nuclear envelope. The nuclear lamina is thought to provide an architectural framework for the attachment of both nuclear membranes and nuclear pore complexes. The association of lamins with the scaffold fraction has led to the idea that the lamina anchors interphase chromatin to the nuclear envelope and thereby influences higher order chromosome structure (Fig. 2.21).

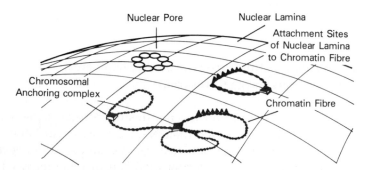

Figure 2.21. Potential association of the chromatin fiber with the nuclear lamina.

The possible interaction between the nuclear lamina and chromatin is supported by electron microscopic studies which have indicated that chromatin is in intimate contact with the inner surface of the nuclear envelope. Biochemical experiments indicate that solubilization of nuclear membranes with Triton X-100 leaves the lamina in association with chromatin. More recently Gerace and colleagues have made use of an *in vitro* nuclear assembly system (Section 3.4.4) to examine the reassembly of nuclear membranes and lamina around mitotic chromosomes. They found that lamins A and C specifically bind to mitotic chromosome surfaces and assemble a supramolecular structure. In fact, lamin assembly is facilitated by the presence of chromosomes, lending strong support to the hypothesis that the chromosome structure influences lamina structure and nuclear reassembly (Glass and Gerace, 1990).

A second set of proteins found in the scaffold fraction includes Sc (scaffold proteins) I, II and III (Lewis and Laemmli, 1982). The function of Sc II and III is unknown; however, Sc I (170 kDa) is now known to be topoisomerase II. The enzymatic activity of topoisomerase II is to pass DNA strands through one another. It can cause a double-strand break in DNA and rejoin it. (Topoisomerase I can only introduce a single strand break in the double helix, see Table 2.1.) Antibodies to topoisomerase II allowed the demonstration that the protein is an integral component of mitotic chromosomes (Earnshaw *et al.*, 1985; Earnshaw and Heck, 1985). Moreover, the efficiency of recovery of total cellular topoisomerase II in the scaffold fraction (> 70%) makes it unlikely that the association with the scaffold fraction is accidental. Consistent with the current view of the folding of the chromatin fiber (Section 2.4.1), immunolocalization data show that topoisomerase II is found in a large number of discrete foci, scattered throughout the axial region of chromosomes. These foci are very uniform in size suggesting that they represent discrete structural complexes. Each is believed to be an anchoring complex to which chromatin loops are attached. The presence of topoisomerase II in these complexes can be rationalized by the necessity of unravelling DNA knots and tangles that will inevitably be generated during processive enzymatic processes such as replication and transcription (Sections 4.3.1 and 4.3.4). In fact, if topoisomerase II is inactivated *in vivo* by mutation, the mutant cells die because they cannot separate their chromosomes at the end of mitosis (DiNardo *et al.*, 1984).

It has been suggested that the nuclear matrix or scaffold contains specific DNA sequences known as matrix or scaffold attachment regions (MARs and SARs). Topoisomerase II and histone H1 are believed to bind preferentially to these AT-rich DNA sequences (Adachi *et al.*, 1989; Izaurralde *et al.*, 1989). The significance of the presence of residual DNA in scaffold or matrix preparations is controversial since exchange with free DNA sequences in solution can occur during preparation (Jackson *et al.*, 1990). Substantial evidence exists to suggest that at some time in the cell cycle all DNA in a chromosome will have some attachment to the nuclear matrix, as DNA replication appears to occur with components attached to such a scaffold (Jackson and Cook, 1986; Cook, 1991).

More recently, the development of *in vitro* nuclear assembly systems has also allowed a direct demonstration of an essential role for topoisomerase II in the chromosome condensation process. These studies use embryonic chicken erythroid cells that provide a source of nuclei from a single lineage that varies in topoisomerase II content (Heck and Earnshaw, 1986). The topoisomerase II content of these nuclei changes dramatically during development, with the bulk of the enzyme being lost when mitosis ceases in mature erythrocytes. Chromosome condensation in both mammalian cell and *Xenopus* extract systems is found to be closely correlated with the level of topoisomerase II present in both the erythrocyte nucleus and the egg extract (Adachi *et al.*, 1991). Although the essential role of topoisomerase II in unravelling intertwined DNA molecules within the chromosome is established, it does not seem to have an essential structural role as a 'building block' of the chromosome. Instead the enzyme is a very useful marker for the anchoring sites of the chromatin fiber into loops (Section 2.4.1).

A third set of proteins within the chromosome scaffold has been identified primarily through the use of autoantibodies from patients with rheumatic disease. These include the INCENP (inner centromere proteins) and the CENP-A, B and C proteins. All of these molecules are now known to associate with the centromere. The centromere is the region of the mitotic chromosome that participates in chromosome movement. The mitotic spindle attaches to a specialized structure at the centromere known as the kinetochore. The motor responsible for the movement of the chromosomes towards the spindle poles during mitosis is also located here. The INCENPs (135–150 kDa) bind tightly to mitotic chromosomes between sister chromatids at the centromere and wherever else the two chromatids are in close contact. Furthermore, the INCENPs are released from a chromosome at the beginning of chromosome separation during mitosis (anaphase), suggesting that they may have a role in regulating sister chromatid pairing (Pluta *et al.*, 1990).

The organization of CENP-A, B and C in the centromere is understood in some detail (Fig. 2.22). The centromere can be broken up into

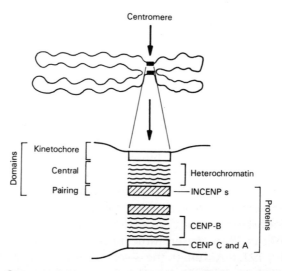

Figure 2.22. Structural domains and protein localization in the centromere of a chromosome.

three distinct structures: the kinetochore, the central and pairing domains. Immunological staining reveals that the INCENPs are in the pairing domain. The central domain contains dense chromatin known as constitutive heterochromatin (Section 2.5.5). The kineto-chore appears to be anchored to this heterochromatin. The DNA within this heterochromatin is composed primarily of various families of repetitive DNA (satellite DNAs).

The α-satellite family of DNA sequences (comprising 5% of the human genome) is probably present at the centromeres of all human chromosomes. The basic repeat, 171 bp in length, occurs in large arrays of up to 3×10^6 long. Nucleosomes have long been known to be rotationally positioned on α-satellite DNA, however the histone core is believed to adopt several distinct translational positions (Simpson, 1991; Section 2.2.5). An important point is that a single nucleosome is believed to exist on every 171 bp repeat. Two non-histone chromosomal proteins are also known to associate with α-satellite DNA: a 10 kDa protein, HMG-I (high mobility group protein, Section 2.5.6), and CENP-B. HMG-I, which binds to α-satellite DNA specifically *in vitro* (Solomon *et al.*, 1986), binds in the minor groove of the double helix recognizing runs of six or more AT base pairs, most probably through certain secondary structural features of the DNA molecule (Section 2.1.1). CENP-B binds to α-satellite DNA specifically recognizing a 17 bp DNA sequence (5′-CTTCGTTGGAAACGGGA-3′) present in a subset of α-satellite DNA repeats (Masumoto *et al.*, 1989). CENP-B contains anionic regions rich in aspartic and glutamic acid residues. These are characteristic of many proteins that interact with chromatin.

CENP-A is especially interesting since it shares homology with core histone H3 (Section 2.5.1). This raises the possibility that some special nucleosomal structure is present at the centromere, in addition to the presence of positioned nucleosomes. The sublocalization of CENP-A or CENP-C within the centromere is unknown. However, CENP-C is only present in centromeres actively engaged in mitosis.

Aside from the centromere, other specialized chromosomal structures associated with non-histone proteins exist. Among the best studied are the telomeres, the ends of eukaryotic chromosomes (Zakian, 1989). These specialized structures protect the chromosomes from exonucleolytic attack, prevent end-to-end fusion of the chromosomes and promote the complete replication of the ends of the linear DNA molecules present in chromosomes. Telomeric DNA has been studied in many organisms including *Tetrahymena* and yeast. Chromosomes of *Saccharomyces cerevisiae* terminate in 250–650 bp of the simple repetitive DNA sequence $C_{2-3}A(CA)_{1-6}$. These are binding sites for a non-histone protein known as RAP 1 (Conrad *et al.*, 1990). Although some telomeres have a nucleosomal organization (Gottsch-ling and Cech, 1984), those of a number of species including *Tetrahy-mena* and yeast display a protein-dependent protection from nu-cleases in which the size of the protected structure differs from that expected for nucleosomes (< 140 bp). This suggests either non-histone protein complexes exist in a regular array or that modified nucleosomes exist at the telomere. The RAP 1 protein may be a participant in the non-histone protein–DNA complexes at the chromosome ends. RAP 1 interacts with yeast telomeres *in vivo* and has been shown to be important for maintenance of telomere length. The protein is abundant (> 4000 copies per cell) and fractionates with the nuclear scaffold. Interestingly it is capable of deforming DNA into curved structures and has some sequence similarity to a known mam-malian transcription/replication factor, NF1/CTF. Later we will discuss the influence of proximity to the telomeres as an example of the effect of chromosomal context on gene expression (Section 2.5.5).

To this point we have focused attention on the structure and properties of normal chromosomes such as might be found in any somatic cell. Much of our understanding of chromosomal structure has followed the observation of unusual chromosomes present in particular organisms or tissues during development. These include amphibian lampbrush chromosomes and the polytene chromosomes of insects.

Summary

Biochemical fractionation and immunological localization have defined components of the nuclear scaffold. Proteins involved in the large-scale architecture of chromosomes and the nucleus include the lamins, topoisomerase II, components of the centromere and telomere. These proteins may recognize DNA, other non-histone proteins or histones within the chromatin fiber or some special nucleosome structure.

2.4.3 Lampbrush and polytene chromosomes

Lampbrush chromosomes are found in the oocytes of many animals (Callan, 1986). They contain very transcriptionally active DNA, where loops of DNA emerging from an apparently continuous chromosomal axis are coated with RNA polymerase. Each RNA polymerase is attached to nascent RNA and associated proteins generating a visible 'brush-like' appearance (Fig. 2.23). The axes of lampbrush chromosomes from which the loops project consist visually of linear arrays of compacted beads, known as chromomeres. DNA is concentrated in the chromomeres which represent compacted regions of chromatin. The axis along which the chromomeres exist consists of two distinct strands of chromatin. Gall and co-workers measured the kinetics of lampbrush chromosome breakage by DNaseI. These experiments led to the important conclusion that there were two DNA duplexes along the chromomere axis, but only one duplex in each transcriptionally active loop of the chromosome that emerges from that chromomere axis. Distinctive loops can be recognized at invariant positions of the chromosomes, and depend upon the DNA sequence contained within each loop (Callan *et al.*, 1987). Each loop may contain several-transcription units and range in size up to 100 kb. Nucleosomes can clearly be seen within active transcription units, especially where RNA polymerases are not so densely packed. An implication of this observation is that nucleosomal structures must be able to reform very rapidly following the passage of RNA polymerase (Sections 4.3.4 and 4.3.5).

Chromomeres occur where there are long regions of inactive chromatin that are compacted into higher order structures, in some respects resembling superbeads (Section 2.3.2). Several groups have prepared antibodies against amphibian oocyte nuclear proteins and have used oocyte sections or isolated lampbrush chromosomes for intranuclear localization. The large size of the chromosomes, their ease of manipulation and the wealth of morphological detail make them ideal for such studies. Actin, histone H2B, nucleoplasmin (Section 3.4.3) and RNA binding proteins have all been localized within loops (Roth and Gall, 1987). Actin filaments may be involved in the extension of the lampbrush chromosome loop away from the chromomere axis.

The polytene chromosomes of insects have had an equivalent utility for the investigation of chromosomal structure and function. The secretory cells of certain insect larvae have an enormous biological requirement for the accumulation of specific mRNAs. In order to fulfill these demands the cells grow to a large size, carrying out multiple cycles of DNA synthesis without cell division. Eventually these giant cells contain over a thousand times the quantity of DNA found in a normal nucleus. The chromosomes within these cells are immense. All of the homologous pairs of chromosomes remain side by side, forming a single giant chromosome. Like lampbrush chromosomes, it is possible to isolate polytene chromosomes under physiological ionic conditions with higher order structure preserved.

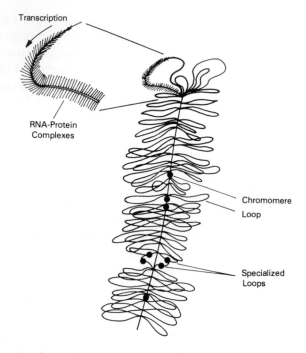

Figure 2.23.   A schematic representation of a lampbrush chromosome with an expanded view of a portion of a single loop.

This loop is shown with attached RNA–protein complexes which grow larger as transcription proceeds. Occasionally the chromatin in the axis of the chromosome is condensed to form a chromomere. There are also specialized loops present with unusual morphologies. These generally synthesize special complexes of RNA and protein.

Immunological staining has revealed the specific distribution of several non-histone proteins including RNA polymerase II and proteins apparently responsible for the generation of inactive heterochromatin (HP1, see Section 2.5.5).

More detailed analysis, especially of the Balbiani ring genes in the polytene larval salivary glands of *Chironomus tentans*, have yielded much information. Balbiani ring genes (37 kb in length) constitute well-defined transcription units, with the first engaged RNA polymerase molecule representing the approximate start site of transcription and the last polymerase the site of termination (Bjorkroth *et al.*, 1988; Ericsson *et al.*, 1990). Within each visible puff or expanded region of transcriptionally active chromatin, each Balbiani ring gene forms a loop (rather like in the lampbrush chromosome). The chromatin axis is fully extended during transcription, while after the completion of transcription, it coils into a 30 nm chromatin fiber and is finally packaged into a supercoiled loop of the chromatin fiber. Upstream of the start site of transcription is a region free of nucleo-some structures, presumably corresponding to the promoter, while further upstream and downstream of the gene are compacted chro-matin fibers. Recent experiments using Balbiani ring chromatin have indicated that some histone H1 remains on chromatin even while it is actively transcribed. Thus by using immunological staining and keen observation a rather complete picture of the transcription process can be established confirming many biochemical deductions.

Summary
Visualization of large lampbrush and polytene chromosomes has provided morphological detail concerning many nuclear processes, especially transcription.

2.5 MODULATION OF CHROMOSOMAL STRUCTURE

Chromosomal structure is not inert. Studies of the molecular mechan-isms regulating the condensation and decondensation of chromo-somes during the cell cycle demonstrate that gross morphological changes in chromatin structure are driven through reversible modifi-cation of chromosomal proteins. This section concerns the structural consequences of chromosomal protein modification, including the significance of histone variants and high mobility group proteins.

2.5.1 Histone variants

Histone genes are invariably present in multiple copies, the level of reiteration varying from two copies of each gene in yeast (excluding histone H1) to several hundred-fold in sea urchin. In some organisms, most notably the sea urchin, distinct batteries of different forms of core histone genes are transcribed at precise periods in development (Poccia, 1986). Such variants are particularly prevalent in gametes, where the key function of the histone is to compact the DNA. Meta-bolic activities involving DNA are generally inhibited in spermatozoa and it is likely that DNA can be compacted in a wide variety of ways. This is because once it is compacted no significant process involving DNA will occur until fertilization (Section 2.5.4). The sea urchin has specialized histone H2A variants (five in all) which have different expression profiles for the cleavage, blastula and gastrula stages of embryogenesis, four different histone H2B variants are also developmentally regulated (Table 2.2). Interestingly, a single form of both histones H3 and H4 is present throughout development, reflecting the central role of these histones in nucleosome structure (Section 2.2.4) and chromatin assembly (Sections 3.4.1–3.4.3). Variations in the primary structure of histones H2A and H2B are likely to alter the compaction of DNA into both the nucleosome and the chromatin fiber. This could be due either to a direct effect on nucleosome structure or an altered binding of histone H1 to the nucleosome core particle (Section 2.3.1). Differences have been observed in the stability of nucleosomes containing early and late sea urchin histones (Simpson and Bergman, 1980).

As might be expected the form of linker histone varies during sea urchin development and during that of many other organisms, often in a tissue-specific way (Wolffe, 1991a). At least six variants of histone H1 are present during sea urchin embryogenesis. Like histone H2A, there are distinct cleavage stage, blastula and gastrula proteins. Interestingly, these latter proteins do not contain short peptide sequences known as 'SPKK motifs' (see Section 2.5.3). These sequences are found multiple times in the tails of linker histones and are the sites of phosphorylation by the major mitotic kinase in the cell (called the cdc 2 or MPF kinase) (Section 2.5.3). Hence the gastrula form of histone H1 in the sea urchin cannot be phosphorylated by this particular kinase. Once again a specific role for the different histone H1 variants has not yet been established, but it is clear that the chromatin of the early embryo is less compacted than that of the gastrula (Longo, 1972) and may, therefore, be more accessible to *trans*-acting factors and more easily replicated. It is also possible that the synthesis of histone H1 variants that cannot be phosphorylated by the MPF protein kinase during embryogenesis, reflects a change in the mechanism by which chromosome structure is regulated during the cell cycle (Section 2.5.3).

Histone H1 variants are also expressed during the development of *Xenopus* with interesting consequences for the regulation of differential gene expression. An embryonic form of histone H1 exists (called B4) which is significantly divergent from normal somatic forms of the protein (Smith *et al.*, 1988). This protein appears to be present in the large chromosomes of the cleavage stage *Xenopus* embryo. The form of histone H1 normally present in *Xenopus* somatic cells is very deficient in *Xenopus* eggs (Wolffe, 1989a, b). However, large stores of maternal mRNA encoding histone H1 are present, the translation of

Table 2.2.   Histone modification in development

Histone	Modification	Variants	Chromatin structure	Functional consequences
H4	Acetylation, prevalent in *Xenopus* and sea urchin eggs and early embryonic chromatin	—	Weakens constraint of linker DNA, dissociates H1 Facilitates nucleosome assembly	30 nm fiber destabilized, linker DNA accessible to *trans*-acting factors Facilitates nuclear division
H3	Acetylation (as above)	—	(as above)	(as above)
H2A	—	7 in sea urchin developmentally regulated	Early embryonic forms hinder chromatin compaction	(as above)
H2B	—	4 in sea urchin developmentally regulated	(as above)	(as above)
		Sperm H2B has extended N-terminal tail	Stabilizes nucleosomes and constrains linker DNA	Sperm chromatin rendered inaccessible to *trans*-acting factors and functionally inert
H1	Phospho-rylation	—	Weakens constraint of linker DNA, creates a paradox, since linker DNA is more accessible, but chromosomes condense	Linker DNA accessible to *trans*-acting factors, transcription facilitated
	—	H5 in chicken	Compacts erythrocyte chromatin	DNA made inaccessible to *trans*-acting factors and functionally inert
	—	2 in *Xenopus*	Early embryonic chromatin is less compact, gastrula chromatin more compacted	Accessible to *trans*-acting factors Non-essential genes, such as oocyte 5S RNA genes are repressed. Cell cycle becomes longer
		6 in sea urchin, early embryonic forms have SPKK motifs, late embryonic forms do not	Early embryonic chromatin is not compacted	Chromatin becomes more stable as embryogenesis progresses

somatic histone H1 mRNA being developmentally regulated (Woodland *et al.*, 1979). Following fertilization somatic histone H1 protein synthesis progressively increases until at gastrulation the normal ratio of one histone H1 molecule per nucleosome is present. Concomitant with the appearance of somatic histone H1 in chromatin, the general accessibility of DNA to RNA polymerases decreases and particular families of genes transcribed by RNA polymerase III are repressed (Section 4.2.1). Linker histone variants direct the repression of metabolic activity involving DNA during spermatogenesis in several organisms (Section 2.5.4) and during the heterochromatinization of mature erythrocyte nuclei in the chicken (Section 2.5.5). Clearly the presence of histone variants correlates with significant changes in the transcription of either specific genes or the entire genome.

Summary

Histone variants are especially prevalent during the development of lower vertebrates. These organisms generally have large eggs and have distinct developmental stages with rapid cellular proliferation prior to the onset of transcription. Histone variants found in early developmental stages may serve to compact DNA less tightly, facilitating rapid nuclear division, DNA replication and access to *trans*-acting factors.

2.5.2 Post-translational modification of core histones

Core histones undergo two major post-translational modifications: acetylation and phosphorylation. Both have been the subject of intense interest. Acetylation of the four core histones occurs in all animal and plant species examined (Csordas, 1990). The sites of modification are the lysine residues of the positively charged N-terminal tails (Section 2.1.2), where each acetate group added to a histone reduces its net positive charge by 1. The number of acetylated lysine residues per histone molecule is determined by an equilibrium between histone acetylases and deacetylases. Two populations of acetylated histone appear to exist in a particular cell nucleus. For example, in the embryonic chicken erythrocyte, 30% of core histones are stably acetylated while the acetylation status of about 2% changes rapidly. The pattern of specific lysine residues in the histone tails that are acetylated varies between different species. This non-random usage suggests that some sequence specificity exists for the relevant acetylases and deacetylases (Turner, 1991).

The influence of acetylation of the histone tails on chromatin structure is not well-defined. Hyperacetylation of the histone tails leads to subtle changes in nucleosome conformation (Oliva *et al.*, 1990). However, it appears that the most significant consequences are for protein–protein interactions, either between nucleosomes, with histone H1 or with non-histone proteins. The N-terminal tails of the core histones are accessible to trypsin suggesting that they are exposed on the outside of the nucleosomal core particle (Section 2.2.4). Chemical cross-linking experiments have shown that weak interactions can occur between the N-terminal domain of H2B and linker DNA, although this is a special sperm H2B with a particularly long N-terminal tail (Bavykin *et al.*, 1990). The N-termini of histones H3 and H4 also interact with core-particle DNA. High-resolution protein nuclear magnetic resonance data indicate association of the N-terminal tails of histones H3 and H4 with DNA in the nucleosome core particle at physiological ionic strength (< 0.3 M NaCl). In contrast, the tails of the normal somatic variants of histones H2A and H2B are mobile at all ionic strengths examined (Cary *et al.*, 1978). The weak interaction of the histone tails with DNA in the nucleosome is reflected in the lack of structural change in the organization of DNA and in the integrity of the nucleosome following their proteolytic removal (Hayes *et al.*, 1991b; Ausio *et al.*, 1989).

Induction of histone hyperacetylation with the deacetylase inhibitor sodium butyrate increases the accessibility of DNA in chromatin to DNaseI. High levels of histone acetylation improve chromatin solubility suggesting a reduced tendency to aggregate (Perry and Chalkley, 1981). The association of histone H1 with chromatin is also reduced, indicating an influence on higher order structure (Reeves *et al.*, 1985). This would be consistent with the possible positioning of the N-terminal tails of histones H3 and H4 adjacent to the histone H1 binding site (Glotov *et al.*, 1978; Section 2.2.4), and the effect of acetylation of the histones tails of H3 and H4 on DNA topology. Bradbury and colleagues have quantitated the number of superhelical turns introduced into DNA per nucleosome, dependent on whether or not the core histones are or are not acetylated. They find that fewer superhelical turns are introduced by a population of nucleosomes when the core histones are acetylated (Norton *et al.*, 1990). As detailed analysis reveals that the helical periodicity of DNA in the nucleosome is unchanged by histone acetylation, by elimination, this suggests that the path of DNA between nucleosomes or writhe of DNA on the histone core is affected by histone acetylation. An influence of the histone tails on the structure of linker DNA could explain their potential involvement in the assembly *in vitro* of higher order chromatin structures.

The generation of antibodies against acetylated histones has allowed a number of general correlations to be made concerning possible functional roles of histone acetylation. In *Tetrahymena*, *Xenopus* and man there is excellent evidence for the presence of acetylated histones in chromatin immediately following replication (deposition-related). Histone acetylation plays an important role in facilitating chromatin assembly (see Section 3.3 for discussion). There is also a strong correlation between histone acetylation and the transcriptional activity of chromatin (Gorovsky *et al.*, 1973). In higher eukaryotes acetylation of histone H4 increases in the transcriptionally inactive chicken erythrocyte nucleus following the fusion of the erythrocyte with a transcriptionally active cultured cell to form a heterokaryon (Turner, 1991; Section 3.1.2). Histone acetylation has been shown to be particularly prevalent over specific genes that are actively transcribed in erythrocytes (Hebbes *et al.*, 1988). Immunolabeling of polytene chromosomes in *Chironomus* and *Drosophila* also reveal a non-random distribution of histone H4 acetylation correlating with transcriptional activity. Therefore, several independent experimental approaches have shown that actively transcribed genes are selectively enriched in acetylated histones (Section 4.3.4).

The second type of core histone modification to receive extensive experimental study is phosphorylation. Histone H3 is rapidly phosphorylated on serine residues within its basic N-terminal domain, when extracellular signals such as growth factors or phorbol esters stimulate quiescent cells to proliferate (Mahadevan *et al.*, 1991). The basic N-terminal domain of histone H3, like that of histone H4, may interact with the ends of DNA in the nucleosomal core particle and therefore perhaps with histone H1 (Glotov *et al.*, 1978). Several studies have suggested a change in nucleosomal conformation or higher order structure within the chromatin of the proto-oncogenes *c-fos* and *c-jun* following their rapid induction to high levels of transcriptional activity by phorbol esters (Chen and Allfrey, 1987; Chen *et al.*, 1990). DNaseI sensitivity of chromatin rapidly increases and proteins with exposed sulfhydryl groups accumulate on the proto-oncogene chromatin. The proteins containing exposed sulfhydryl groups include both non-histone proteins such as RNA polymerase and molecules of histone H3 with exposed cysteine residues. The histone H3 cysteine residues, the only ones in the nucleosome, are normally buried within the particle. Exposure of the sulfhydryl groups might imply a major disruption of nucleosome structure. For example, the dissociation of an H2A/H2B dimer might allow access from solution to this region of histone H3. Phosphorylation and acetylation of histone H3 might act in concert to cause these changes (Section 4.3.4).

A different type of function for core histone phosphorylation is provided by histone H2A. Here, histone phosphorylation appears to be involved in directing the spacing of nucleosomes along DNA (Section 3.4.2).

Core histones are also methylated on their lysine residues with no clear functional consequences. Some evidence suggests that ADP-ribosylation of core histones may lead to localized unfolding of the chromatin fiber. ADP-ribosylation may play a particularly important role in DNA repair (Section 4.3.3). Here the disruption of chromatin structure cannot rely on the processive enzyme complexes involved in DNA replication or transcription. The synthesis of long, negatively charged chains of ADP-ribose may well facilitate a partial disruption of nucleosomes, presumably by exchange of histones to this competitor polyanion. Histone H2B and especially H2A can also be modified by addition of the small protein ubiquitin. Ubiquitin has been found to participate in regulating protein degradation. The protein is covalently attached, via an ATP-dependent reaction, to a protein to be targeted for proteolysis. The significance of ubiquitination of histones H2A and H2B for chromatin structure is unknown (van Holde, 1989).

Summary

Core histone acetylation has important consequences for the organization of DNA in a nucleosome, loosening interactions at the periphery of the structure and probably facilitating the dissociation of histone H1. Core histone phosphorylation may also have important structural consequences for nucleosome structure and spacing.

2.5.3 Linker histone phosphorylation

The most studied post-translational modification of chromatin is the reversible phosphorylation of histone H1. This modification varies through the mitotic cell cycle (Fig. 2.24). Studies of both the slime mold *Physarum* and mammalian cells in culture show that phosphorylation of H1 is highest in rapidly dividing cells and decreases in non-proliferating cells; levels of histone H1 phosphorylation are lowest in G1 and rise during S phase and mitosis. During mitosis, phosphorylation of histone H1 peaks at metaphase when chromosomes are at their most condensed. These results have led to the suggestion that a

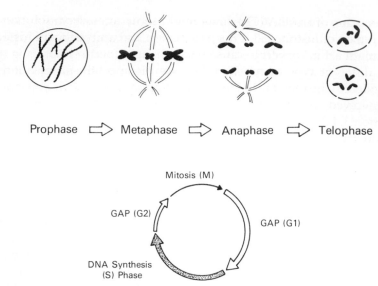

Prophase ⟹ Metaphase ⟹ Anaphase ⟹ Telophase

Mitosis (M)

GAP (G2)

GAP (G1)

DNA Synthesis
(S) Phase

Figure 2.24.   The eukaryotic cell cycle.
The upper panel shows stages of mitosis and their characteristic chromosome morphologies. The lower panel shows the entire cell cycle.

causal relationship exists between histone H1 phosphorylation and
chromosome compaction (Bradbury *et al.*, 1974).

Histone H1 consists of a globular central domain flanked by lysine-
rich highly basic N-terminal and C-terminal tails (Section 2.3.1). The
globular domain interacts with DNA in contact with the core histones, whereas the tails bind to linker DNA. Phosphorylation of the
histone H1 tails occurs predominantly at conserved (S/T P-X-K/R,
serine/threonine, proline, any amino acid, lysine/arginine) motifs of
which several exist along the charged tail regions (Churchill and
Travers, 1991). It might be expected that the neutralization of positive
charge on the tails would weaken the interaction of histone H1 with
the linker DNA. This might be a prerequisite for chromosome condensation, but is also paradoxical since the association of histone H1
with linker DNA is thought to direct the folding of nucleosomal
arrays into the chromatin fiber (Section 2.3.2).

In order to resolve this paradox it has been useful to examine
systems in which mitosis and chromosome compaction are uncoupled (Roth and Allis, 1992). *Tetrahymena* is a ciliated protozoan in
which two distinct nuclei exist differing in structure, function and
mitotic behavior. The somatic macronucleus is responsible for maintaining cell growth, is transcriptionally active and divides amitotically

without any apparent condensation of chromatin. Surprisingly, however, macronuclear H1 phosphorylation is controlled through a kinase (the cdc 2 or MPF activity) that is similar to that regulating the cell cycle in normal mammalian cells. Nevertheless the activity of this kinase and chromosome condensation can be uncoupled. The phosphorylation state of macronuclear histone H1 is highly dependent upon cell growth conditions. If the cells are starved, growth ceases and histone H1 is moderately dephosphorylated. More significantly, during conjugation the macronucleus becomes completely inert, chromatin condenses and histone H1 is completely dephosphorylated (Lin *et al.*, 1991). Thus phosphorylation of H1 is inversely related to chromosome condensation. In contrast to the macronucleus, the germline micronucleus which is responsible for the sexual cycle is normally transcriptionally silent and undergoes a normal mitotic cycle including the formation of mitotic chromosomes (Gorovsky, 1986).

Several very useful studies correlating nuclear function and histone modification have been carried out comparing these two nuclei of *Tetrahymena* within the same cell – a natural heterokaryon (Section 3.1.2). Briefly, special variants of histone H2A are present in the transcriptionally active macronucleus but absent from the micronucleus. Histones are also more extensively acetylated in the macronucleus (Section 2.5.2). The association of the linker histone H1 with chromatin in the macronucleus also decreases with transcriptional activity. This macronuclear histone H1 is highly phosphorylated during vegetative growth. In micronuclei, macronuclear histone H1 is replaced by four specialized linker histone polypeptides (Allis and Gorovsky, 1981; Roth *et al.*, 1988).

A second system in which linker histone phosphorylation can be uncoupled from mitosis concerns the function of the specialized linker histone H5 during development of chicken erythroid cells. During the final stages of chicken erythrocyte development, the nucleus is condensed into inactive heterochromatin due in part to the appearance of histone H5 (Section 2.5.5). Topoisomerase II also becomes much reduced during this differentiative process (Section 2.4.2). Newly synthesized histone H5 is highly phosphorylated, but when the erythrocyte chromatin is becoming condensed, histone H5 is quantitatively dephosphorylated. Hence once again dephosphorylation of a linker histone correlates with chromatin compaction (Aubert *et al.*, 1991). That these two events are directly linked receives further support from experiments in which the gene for histone H5 was expressed in fibroblasts. This specialized linker histone would not normally be found in these cells. The accumulation of histone H5 in the fibroblasts inhibited cell growth concomitant with chromatin compaction. Under these circumstances histone H5 was not phosphorylated. Introduction of the protein into transformed cells led to phosphorylation of histone H5. In this case nuclear condensation did not occur and the cells continued to grow and divide. Phosphorylation of the linker histone clearly prevents chromosome folding as might be expected from biophysical analysis (Section 2.3.1).

The final example of a correlation between linker histone phosphorylation and chromatin compaction concerns sea urchin spermatogenesis. Here dephosphorylation of a sperm-specific histone H1 correlates with chromatin condensation (Green and Poccia, 1985). Following fertilization, sperm histone H1 is phosphorylated in parallel with decondensation of the sperm pronucleus (Section 3.4.4). In all of these examples we see that histone H1 dephosphorylation correlates with chromosome compaction. The coordinate phosphorylation of non-histone proteins at the same time as histone H1 during the cell cycle seems more likely to regulate the compaction of chromatin during mitosis (see below).

Phosphorylation of histone H1 has been shown directly to weaken interaction of the basic tails of the protein to DNA; surprisingly these changes influence the binding of the protein to chromatin even more than to DNA (Hill *et al.*, 1991). Perhaps phosphorylation of histone H1 is required to weaken the interaction of the linker histone with chromatin and thereby 'loosen' the chromatin fiber to allow other *trans*-acting factors required for gross changes in chromosomal architecture to bind to DNA or the fiber itself (Section 4.2.2).

Characterization of the major mitotic kinase (cdc 2 or MPF) in eukaryotic cells has allowed many of the nuclear events driven by phosphorylation to be defined (Dunphy and Newport, 1988). In the course of these studies it became clear that MPF was similar if not identical to the major histone H1 kinase in eukaryotic cells. During mitosis in higher eukaryotes, MPF induces the ultrastructural changes required for cell division; these include nuclear envelope disassembly (nuclear membrane and lamina), chromatin condensation and construction of the mitotic spindle (Fig. 2.24). Although histone H1 becomes hyperphosphorylated during mitosis, it is clearly not the only substrate for MPF during the cell cycle. Newport, Gerace and colleagues have shown that disassembly of the nucleus, nuclear membrane vesicularization, lamin disassembly and chromosome condensation are all independent processes (Ohaviano and Gerace, 1985; Newport and Spann, 1987; Newport *et al.*, 1990). The disassembly of the nuclear lamina appears to be driven by phosphorylation. It is quite possible that phosphorylation of the other proteins found in the nuclear scaffold fraction inducting Sc I (topoisomerase II), Sc II and Sc III may influence chromatin and chromosome folding.

Summary
Linker histone phosphorylation in systems uncoupled from mitosis leads to decondensation of chromatin. Consequently the increase in phosphorylation during mitosis is paradoxical and of unresolved functional significance.

2.5.4 Remodeling of chromatin during spermatogenesis

Histones represent only one way of packaging DNA such that it can fit into the volume of a nucleus. There are many possible ways of rendering DNA compact in a reversible fashion. It is a measure of the major role of histone structure in many other nuclear processes (Sections 4.2 and 4.3) that histones have been so conserved through evolution. Perhaps the best example of the reversible compaction of DNA by multiple pathways concerns the condensation of DNA into sperm nuclei during spermatogenesis. This provides an excellent example of roles for histone variants, post-translational modification of histones and non-histone proteins in regulating chromosome structure and function.

The easy availability of pure populations of spermatozoa led to the early realization that the types of proteins in the sperm nucleus could vary greatly in different organisms. These proteins have been divided into certain classes. One class, the protamines, comes in two types: one type is rich in polyarginine tracts (4–6 residues), punctuated with proline, and potentially phosphorylatable serine and threonine residues; the other type is rich in cysteine. Both types of protamine are small (3000–5000 Da) and highly basic. Protamine–DNA complexes often represent the final state of chromatin in fish and mammalian-sperm nuclei. However, during the process of spermatogenesis other proteins can transiently replace the histones (Poccia, 1986).

The transition proteins replace histones during the initial stages of condensation of chromatin in spermiogenesis and are later replaced by protamines which are the only basic nuclear structural proteins in the sperm of most mammals. The transition proteins presumably facilitate the replacement of histones by protamines, although little is known about their specific function. The amino acid sequence of a mouse transition protein (TP 2) suggests two domains of protein structure: an N-terminus which, like the protamines, is rich in cysteine, proline and phosphorylatable serine and threonine, and a highly basic C-terminus rich in arginine (Kleene and Flynn, 1987; Luerssen *et al.*, 1989). The transition proteins are of especial interest because they are the only proteins (together with protamines) that are proven to dissociate histones from DNA in a physiological context (without replication, Section 2.3.2). Several mechanisms have been proposed. First, the electrostatic binding of the transition proteins to DNA should be intermediate between histones and protamines because the concentration of basic amino acids near the C-terminus is greater than that of histones and less than in protamines. Second, the putative DNA binding domain contains aromatic amino acids: phenylalanine and tyrosine. The tyrosine residues in the transition proteins have been shown to intercalate between the bases of DNA, lowering its thermal stability (Singh and Rao, 1987). Intercalation of aromatic amino acids can induce bends or kinks in native DNA, and these bends might alter the path of DNA around the histone core and potentially destabilize the nucleosome.

Special histone variants exist that are specific for the testis. In sea urchin sperm, specific variants exist of histones H2B and histone H1. Both molecules are considerably larger and richer in arginine than their relatives in somatic cells. Most of the extra size is due to N-terminal extensions. These additional amino acids increase the interaction of both histones with linker DNA (Bavykin *et al.*, 1990; Hill and Thomas, 1990). It is possible that they might also further facilitate the stabilization and condensation of chromatin by creating *inter*-strand linkages between chromatin fibers. Phosphorylation of the histone H1 tails regulates their interaction with DNA (Hill *et al.*, 1991). As chromatin is compacted the tails are dephosphorylated (Section 2.5.3). Testis-specific histone variants also exist in mammals. Variants of histone H1, histones H2A and H2B accumulate during meiotic prophase. As all of these transitions in chromatin structure occur after replication, the movement of a processive enzyme complex along the DNA duplex is not a prerequisite for the remodeling of chromosomal structure (Sections 3.1.1 and 3.1.2). In the rat, the core histone variants are heavily acetylated prior to their dissociation from DNA, which is driven by the accumulation of the transition proteins. Finally, accumulation of the protamines, whose binding is also regulated by their phosphorylation state, leads to the progressive displacement of the transition proteins and the nucleus is completely condensed. The problem of decondensing the sperm nucleus following its arrival in the cytoplasm of the fertilized egg will be considered later (Section 3.2.1).

Why is sperm chromatin condensed through such a special mechanism? Suggestions include the protection and stability of the genetic material DNA in such structures. DNA in the nucleus of a spermatozoon is much less accessible to nucleases than in a somatic cell. It is also much more stable to physical and chemical perturbants. A concomitant effect of chromatin compaction, as normal histones are replaced in chromosomes, is the suppression of gene activity. It is possible that the compaction of DNA into the sperm nucleus not only renders the DNA inaccessible to RNA polymerases but also erases the developmental history of a chromosome. Specifically the *trans*-acting factors responsible for directing specific events in the nucleus could be displaced. Evidence to support this concept comes from an analysis of DNaseI-hypersensitive sites in chromatin (Section 4.2.4) which are lost in sperm for all genes. DNaseI-hypersensitive sites occur where the chromatin fiber is disrupted by *trans*-acting factors. However, those genes that are constitutively expressed are marked during spermatogenesis by hypomethylation at sites of future hypersensitivity (Groudine and Conkin, 1985).

Summary
Spermatogenesis requires the packaging of DNA into an inert chromatin structure such that DNA can be unfolded rapidly following fertilization. A vast variety of proteins have been found that accomplish this packaging probably because little metabolic activity involving DNA occurs in the sperm nucleus. Histones are removed through modifications such as acetylation and competed away from DNA by very basic proteins such as arginine-rich transition proteins or protamines.

2.5.5 Heterochromatin

Early studies by cytologists led to the realization that some chromosomal regions have properties distinct from the rest of the chromosome (Pardue and Hennig, 1990). Large segments of chromatin were found to be highly condensed and to replicate late in S-phase. Geneticists determined that these chromosomal regions, which they called heterochromatin, did not participate in meiotic recombination. However, heterochromatin does have significant genetic effects. The most common observed influence of heterochromatin formation is the repression of transcription either in heterochromatin itself or in regions of chromatin that lie adjacent to the heterochromatin domain. Two explanations have been offered for this 'position effect'. The first is that special proteins, such as HP1 (see below), exist that cause heterochromatin to adopt its distinct structure and these can 'spill over' into regions of normal chromatin. The second applies only to *Drosophila* and other insects in tissues with polytene chromosomes in that, following placement adjacent to heterochromatin, a gene will undergo fewer rounds of replication than would normally occur. Fewer copies of the gene would cause a concomitant reduction in transcription. Most investigators accept that heterochromatin-specific proteins can diffuse onto normal chromatin and thereby influence gene expression of juxtaposed genes.

A genetic approach to the molecular basis of heterochromatin formation has been to look for mutations in *Drosophila* that enhance or suppress position effects on gene expression. The mutations may occur in the genes encoding the chromosomal proteins involved in forming heterochromatin. Support for this approach comes from the observation that chromosomal deletions that reduce the number of copies of histone genes in *Drosophila* reduce the influence of position effects on gene expression (Moore *et al.*, 1983). Using this approach a gene encoding a non-histone protein (HP1, heterochromatin protein 1) has been identified (James and Elgin, 1986). Mutation of the *HP1* gene reduces position effects on gene expression (Elgin, 1990). *HP1* is preferentially associated with the heterochromatin regions of polytene chromosomes. Other proteins homologous to HP1 include polycomb, which is also chromatin associated and is known from genetic experiments to influence the expression of many genes in normal chromatin (Zink and Paro, 1989). Neither HP1 nor polycomb interact with DNA directly but presumably recognize some aspect of nucleosome or chromatin fiber structure. In contrast a second gene, encoded by the locus *Suvar(3)7* – mutation of which also reduces position effects – encodes a zinc-finger protein (Reuter *et al.*, 1990). The multiple zinc-fingers suggest that the protein may interact with DNA directly at several sites in the chromosome. Immunofluorescent studies similar to those described for *Drosophila* mitotic chromosomes and topoisomerase II (Section 2.4.2). are likely to provide significant mechanistic insights into the molecular basis of heterochromatin formation. A similar genetic analysis in yeast has revealed that genes involved in repression of the silent mating-type locus are also involved in mediating position effects at the telomeres of chromosomes. These include the gene for histones and acetyltransferases (Aparicio *et al.*, 1991).

Two examples of mammalian chromosomal regions that contain heterochromatin are the centromere and the telomere (Section 2.4.2). Heterochromatin at the centromere contains tandemly repeated simple sequence 'satellite' DNA, for example the α-satellite DNA at the human centromere. This α-satellite heterochromatin appears to play a structural role by mediating attachment of the kinetochore. The inactive X-chromosome of female mammals is also heterochromatic. Female mammalian embryos begin development with two active X-chromosomes; however, very early in embryogenesis almost all of the genes on one of the two X-chromosomes become inactivated. This transcriptional inactivation is concomitant with the chromosome both taking on the appearance of heterochromatin and also becoming late replicating. Although in eutherian (placental) mammals the initial choice between inactivation of the maternal or paternal X-chromosome is random, once established in a repressed state the same X-chromosome will be inactivated after every cell division. This is an excellent example of the establishment and maintenance of a chromosomal state of determination (Section 4.3.1). The exact molecular mechanism causing heterochromatinization of the inactive X-chromosome remains unknown.

Occasionally an entire nucleus will become heterochromatinized, one example being the inactivation of the erythrocyte nucleus in chicken. Here the special linker histone H5 accumulates, which represses transcription and compacts nucleosomal arrays very effectively. Histone H5 is more arginine-rich than the normal linker histone H1 found in somatic cells. It is this increase in arginine content that probably strengthens its interaction with DNA and stabilizes chromatin structure. Many important experiments have been carried out using histone H5 (Sections 2.5.3 and 3.1.2).

Summary
Several different proteins have been found that mediate the folding of the chromatin fiber into a more compacted and inert state known as heterochromatin. Heterochromatin is important because its formation influences the transcription of genes both within and adjacent to it. This influence is believed to occur through lateral diffusion of the proteins responsible for the additional compaction or stabilization of the chromatin fiber.

2.5.6 Other structural non-histone proteins in the chromosome

Chromatin structure can also be modified by the selective association of abundant non-histone proteins that interact with DNA histone complexes. Primary among these are the high mobility group proteins. Early methodologies for the fractionation of the linker histone H1 employed perchloric acid extraction of chromatin; however, several other proteins were also found to be solubilized during this process. Later it was noticed that extraction of chromatin at moderate ionic strengths (0.35 M NaCl) released similar proteins. The addition of trichloroacetic acid (2%) to these salt-extracted proteins separated them into an insoluble fraction of large proteins (low mobility group, LMG, when molecular size was assayed by gel electrophoresis) and a soluble fraction of small proteins (a group of high mobility proteins during electrophoresis, HMG) (Johns, 1982). Four major proteins are found in the HMG group. These fall into two classes: HMG1 and 2 are one pair of homologous proteins ($\sim$ 29,000 Da); HMG14 and 17 are the other (10,000–12,000 Da). The content of HMG14 and 17 in chromatin may range up to 10% of DNA weight, similar to that of histone H1. In addition there are also several minor HMG proteins, for example HMG-I which binds to the α-satellite sequences in the centromere (Section 2.4.2).

The genes encoding all four of the major HMGs have been cloned. The HMG14 and 17 proteins are highly conserved from man to chicken, certain basic stretches of amino acids being identical. These N-terminal basic regions are believed to interact with nucleosomal DNA. HMG14 and 17 also have an acidic C-terminal tail of unknown function (Srikantha *et al.*, 1988). Both proteins bind selectively to nucleosomal DNA in preference to naked DNA of a comparable length. It appears that two HMG molecules can bind per core particle. One model based on chemical cross-linking suggests that the HMG14 and 17 proteins can interact with DNA where it exits and enters the nucleosome (Shick *et al.*, 1985). Like acetylation or phosphorylation of the core histones interaction at this site is likely to modify histone H1 interaction and hence higher order chromatin structure. Although definitive proof is lacking, considerable circumstantial evidence suggests that HMG14 and 17 are involved in potentiating the transcription of genes (Einck and Bustin, 1985; Section 4.3.4).

Rather more functional information is available concerning the other pair of HMG proteins, 1 and 2. These have attracted a great deal of attention since conserved amino acid sequence motifs within these proteins are also found in transcription factors. HMG1 and 2 have a basic N-terminus and an acidic C-terminus. The basic region contains an 80 amino acid domain that has been found in several transcription factors (Jantzen *et al.*, 1990). Most notable among these is a sequence-specific DNA binding protein (UBF) involved in the regulation of mammalian ribosomal RNA gene transcription (Section 4.1.3). The central region of this HMG box also reveals sequence similarity to the POU transcription factors within a region of these proteins believed important for efficient DNA binding. The UBF transcription factor also has a very acidic C-terminal tail like the HMG1 and 2 proteins. These anionic regions have attracted attention since they are often found in both transcription factors and proteins known to be integrally involved in chromatin structure (Earnshaw, 1987).

Aside from HMG1 and 2, the centromere protein CENP-B has sequences rich in glutamic and aspartic acid residues (Section 2.4.2), chromatin assembly proteins N1/N2 and nucleoplasmin have comparable regions (Section 3.4.2), and topoisomerase I also has a very acidic domain. These anionic regions have been postulated to interact directly with histones, since N1/N2 interacts specifically with histones H3/H4 and nucleoplasmin with H2A/H2B. In fact, HMG1 and 2 will facilitate nucleosome assembly (Bonne-Andrea *et al.*, 1984). The physiological significance of this assembly activity is unknown (Sections 3.4.1 and 3.4.2). It has also been postulated that the acidic regions found within transcription factors might also cause a local destabilization of nucleosome structure, perhaps by competing with DNA for interaction with histones, especially histones H2A/H2B.

Summary
The HMG14 and 17 proteins have a higher affinity for nucleosomal DNA than for naked DNA. They may influence the association of histone H1 with chromatin and consequently the folding of the chromatin fiber.

The HMG1 and 2 proteins interact preferentially with naked DNA. They have a highly conserved DNA binding domain and a domain of acidic amino acids. These anionic regions are commonly found in chromatin-associated proteins.

No definitive function for the HMG proteins in chromatin has yet been determined.

CHAPTER THREE

Chromatin and Nuclear Assembly

Our knowledge of chromatin structure is largely dependent on the analysis of relatively homogeneous populations of both large and small chromosomal fragments. The methodological approach has been to progressively take the chromosome apart and to examine its constituents. However, in order to understand a complex multicomponent structure completely, we must also be able to reassemble it from its constituents. Considerable progress has been made towards understanding the organization of chromosomes and nuclei through attempts to reconstruct them both *in vivo* and *in vitro*.

3.1 INTERACTIONS BETWEEN NUCLEAR STRUCTURE AND CYTOPLASM IN THE LIVING CELL

Much of our understanding of nuclear assembly comes from experiments pioneered on the large eggs and oocytes of the frog *Xenopus laevis*. The large size and easy availability of these eggs and oocytes make them a particularly suitable target for the microinjection of macromolecules (Gurdon, 1974). However, there are significant physiological differences between eggs and oocytes. An oocyte is developing into an egg cell, it is located in the ovary, closely surrounded by several thousand follicle cells and cannot be fertilized. During the six or more months of oogenesis the chromosomes of an oocyte are very actively transcribed, sometimes as lampbrush chromosomes (Section

2.4.3). Fully grown oocytes respond to the hormone gonadotrophin by undergoing maturation. This involves the breakdown of the membrane and lamina surrounding oocyte nucleus (the germinal vesicle), release of the oocyte from the ovary (ovulation), as well as the completion of the first meiotic division with arrest at the second meiotic metaphase. When released from the frog, the egg can be fertilized, and if so will develop very rapidly (3 days) into a free swimming tadpole. This rapid development is due in part to the large stores of nuclear components sequestered in the egg during oogenesis. The great majority of microinjection experiments are carried out on fully grown oocytes, removed from the ovary of a female, or on recently fertilized eggs.

3.1.1 Nuclear transplantation

The first experiments to broadly address directed alterations in nuclear architecture *in vivo* under controlled conditions were those that examined the consequences of introducing the nuclei of somatic cells into *Xenopus* eggs and oocytes. When the nucleus of a single somatic cell is injected into an enucleated egg, this nucleus will promote development of the egg into an embryo and sometimes into a tadpole or adult frog (Gurdon, 1974). However, the more differentiated the cell from which a donor nucleus is taken, the more unlikely it is that correct development will proceed. For example, it has still not yet been possible to obtain a normal adult animal by the transplantation of the nucleus of an adult somatic cell into an egg. These nuclei have, however, yielded swimming tadpoles with functional differentiated cells of most kinds. The significance of these results for our purposes is that they suggest that nuclei from differentiated somatic cells are totipotent (Di Bernardino, 1987). Totipotency requires that whatever genetic mechanisms operate to direct differentiation, they are completely reversible. This has important implications for the role of chromatin structure and *trans*-acting factor–DNA interactions in establishing and maintaining stable states of differentiated gene activity. Although in certain instances cell specialization is coupled to the loss of DNA from the eukaryotic cell or the irreversible rearrangement of DNA sequences, these phenomena are not generally observed (Klobutcher *et al.*, 1984; Hood *et al.*, 1985). Experiments in other systems including heterokaryons, cell hybrids and tumor cells, also point towards the reversibility of the differentiated state under certain conditions (Section 3.1.2).

The pioneering studies on the behavior of somatic nuclei

transplanted into eggs revealed that these nuclei swelled over 60-fold after injection. Furthermore, this change in nuclear volume correlated with the capacity of these nuclei to initiate DNA synthesis (Graham *et al.*, 1966). Subsequent experiments have suggested that the major failure of differentiated cell nuclei to fulfil a complete developmental program after transplantation is a consequence of chromosomal damage. This appears to be due to the premature initiation of DNA replication in chromosomes not appropriately organized for this highly regulated process to proceed correctly (Di Bernardino, 1987).

Early experiments examining specific gene activity within transplanted somatic nuclei, revealed that nucleoli disappeared and previously active ribosomal RNA genes were inactivated (Gurdon and Brown, 1965). Nucleoli are an example of the localization of a particular biosynthetic event, the synthesis of rRNA, to a specific chromosomal structure. The inhibition of ribosomal RNA transcription in eggs clearly demonstrated the capacity of egg cytoplasm to influence nuclear function. As development of the embryo containing the transplanted nucleus proceeds, the ribosomal genes were reactivated and nucleoli reappeared. This influence of cytoplasm on nuclear function could be better understood once it was shown that a considerable movement of proteins from the egg cytoplasm to the somatic nucleus occurred following transplantation (Merriam, 1969). This movement was concomitant with nuclear swelling and with a significant reduction in the amount of heterochromatin within the nucleus. The capacity of nuclei to restructure in this way was much more effective for egg as opposed to oocyte cytoplasm (Gurdon, 1968, 1976). These results were placed into a more physiological context by the observation that *Xenopus* sperm nuclei increase in volume over 50 times in egg cytoplasm within 30 min of fertilization. The relatively slow increase in nuclear volume following transplantation of somatic nuclei into oocytes was shown to be coincident with an increase in transcriptional activity of the nuclei. Unlike eggs, nuclei transplanted into oocytes do not replicate their DNA. Transcriptional activity was reflected in an enlargement of the oocyte nucleoli and enhancement of ribosomal RNA synthesis (Gurdon, 1976).

Both the enlargement of somatic nuclei and their concomitant increase in transcriptional activity were found to be enhanced by rupture of the large oocyte nucleus (the germinal vesicle). This suggested that large stores of nuclear components were present in this enormous nucleus (50 μm diameter). HeLa cell nuclei can enlarge over 500 times in oocytes. Although large quantities of both histone and non-histone protein are taken up by the enlarging nuclei, over 75% of pre-existing nuclear protein was lost (Gurdon *et al.*, 1976).

Together, these observations represent the first evidence for a comprehensive remodeling of nuclear structures by *Xenopus* egg and oocyte cytoplasm.

Summary
The first evidence for a remodeling of chromatin structure with important functional consequences came from the microinjection of somatic nuclei into the oocytes and eggs of amphibians.

3.1.2 Heterokaryons

An approach conceptually related to the reprogramming of somatic nuclei following their introduction into the cytoplasm of a *Xenopus* egg or oocyte, is the fusion of two distinct somatic cells to form a single cell with two different nuclei bathed in a common cytoplasm (a heterokaryon). Early experiments had shown that gene expression in the donor cells could be dramatically changed following formation of a heterokaryon. This work was among the first to suggest the existence of specialized *trans*-acting factors in eukaryotes capable of repressing or inducing the expression of differentiated functions, i.e. regulating differential gene expression (Ephrussi, 1972; Ringertz and Savage, 1976). A common observation was that a gene normally only active in a differentiated cell was inactivated upon fusion with a different differentiated or an undifferentiated cell. Somatic cell hybrids in which the two nuclei of the heterokaryon fuse often lose chromosomes in culture. This type of phenomenon led to the attribution of individual repressive effects to particular chromosomes. Very occasionally gene activation was observed in cell fusion experiments. For example, extensive experiments in heterokaryons have clearly shown that fusion of one differentiated cell (a muscle cell) with a different cell in which muscle genes are not normally expressed (a human amniocyte) leads to the activation of muscle genes in the amniocyte (Blau *et al.*, 1983). This result suggests that factors capable of activating genes can either exchange freely between nuclei or exist in excess within the cytoplasm. The activation of differentiated genes in a non-differentiated cell is rapid (within 2 days) and does not require cell division or DNA replication. This implies that genes can be activated (at some level) without requiring replication events. This result is identical in principle to the activation of genes following the introduction of somatic nuclei into *Xenopus* oocyte cytoplasm (Section 3.1.1).

Experimental results with heterokaryons and *Xenopus* eggs have been interpreted as providing evidence for a continuous regulation of a plastic differentiated state. Implicit in this model is the idea that all genes are continually regulated by *trans*-acting factors that can either activate or repress genes (Blau *et al.*, 1985; Blau and Baltimore, 1991). For certain genes this is clearly true; however, it has also been shown that a considerable remodeling of chromosomal structure occurs in *Xenopus* egg and oocyte cytoplasm. A similar, albeit less impressive, remodeling of chromosomes occurs in heterokaryons. For example, the nuclei of chicken erythrocytes consist predominantly of hetero-chromatin containing the specialized linker histone H5 (Section 2.5.5). In heterokaryons formed by fusion of chicken erythrocytes with proliferating mammalian cells the chicken erythrocyte nuclei once again become transcriptionally active. This process is accompanied by decondensation of chromatin, enlargement of the nucleus and the appearance of nucleoli. Transcription and replication of these nuclei is activated. The enlargement of the chicken erythrocyte nucleus is due to a massive, but selective uptake of mammalian nuclear proteins including RNA polymerases. Histone H5 is partially lost from the chicken erythrocyte nucleus and partially taken up by the mammalian nucleus in the heterokaryon (Ringertz *et al.*, 1985). Histones H2A and H2B also exchange under these circumstances, but not histones H3 and H4. These results might be expected considering the relative affinity of the histones for DNA and their organization in the nucleosome (Sections 2.2.2 and 2.2.4). This reorganization is independent of replication. It is therefore clear that chromosome structure is quite dynamic, with some histones (H1, H2A, H2B) continually exchanging with a free pool of proteins in the cytoplasm.

Several experiments suggest that at physiological ionic strength histone H1 rapidly exchanges into and out of the chromatin fiber (Caron and Thomas, 1981). Presumably this dynamic property of the chromatin fiber and the nucleosome would eventually allow many *trans*-acting factors to gain access to their cognate DNA sequences (Section 4.2.3). An important and unresolved question is whether this access is unlimited or whether access is restricted by chromosomal organization. A quantitative determination of whether the level of transcriptional activity following *de novo* activation of a gene in heterokaryon is identical to the transcription of the same gene in a differentiated cell has not yet been made. Of course in *Xenopus* egg cytoplasm, such equivalent activation does occur in order for correct development to proceed through to the tadpole stage; however, here nuclear reprogramming is more rapid and is likely to be facilitated by DNA replication (Section 4.3.1).

Summary
Heterokaryons are formed following the fusion of two different cells such that two distinct nuclei exist in a common cytoplasm. They have been useful in demonstrating a plasticity in gene expression. Previously active or repressed genes can have their expression states changed following cell fusion. Chromatin structure is revealed to be able to change in a reversible way.

3.2 CHROMATIN ASSEMBLY ON EXOGENOUS DNA *IN VIVO*

Although dramatic changes in nuclear architecture can be visualized following nuclear transplantation or heterokaryon formation, in order to understand the consequences of restructuring chromatin, defined templates within cloned DNA molecules need to be introduced into living cells. The chromatin structure of these defined templates reveals that although most DNA can be assembled into nucleosomes, the organization of particular DNA sequences into nucleosomes can be highly selective.

3.2.1 Chromatin assembly in *Xenopus* eggs and oocytes

In a series of important experiments Woodland and colleagues demonstrated that *Xenopus* eggs acquire a huge store of core histones during the growth of the oocyte within the ovary. This store of 140 ng of core histone is enough to make over 20,000 nuclei (Woodland and Adamson, 1977). These core histones are stored in specialized complexes in the oocyte nucleus (Section 3.4.2). In contrast, somatic histone H1 is severely deficient in eggs (Wolffe, 1989a) and appears to be replaced with an embryonic variant (Smith *et al.*, 1988) with specialized functions (Section 2.5.1).

When purified closed circular viral DNA molecules became available in the 1970s, the *Xenopus* oocyte was used as a living test-tube to investigate what functions were encoded by the viral genome. The first experiments showed that viral DNA (SV40) injected in the oocyte cytoplasm was degraded, whereas viral DNA injected into the oocyte nucleus was maintained in a stable supercoiled state (Wyllie *et al.*, 1977, 1978). Each nucleosome is known to stabilize a single supercoil in a closed circular DNA molecule (Section 2.2.3). Furthermore, the injected DNA was transcribed, and assembled into a nucleoprotein complex whose buoyant density and regularly repeated particulate structure was that expected for chromatin. More extensive analysis revealed that each oocyte nucleus could convert a mass of DNA equivalent to 1000 diploid *Xenopus* nuclei into chromatin. In retrospect this is a surprisingly low efficiency of chromatin assembly considering the large stores of histones in the nucleus. However, this may reflect the fact that oocytes are adapted to the storage of nuclear components, whereas eggs are adapted to the assembly of nuclear structures. Efficient transcription of the viral DNA injected into oocytes occurred only under circumstances in which chromatin was formed. In this respect it is of interest that linear DNA does not form chromatin containing regularly spaced nucleosomes as efficiently as negatively supercoiled DNA in *Xenopus* oocyte nuclei, and is transcribed much less efficiently (Mertz, 1982; Harland *et al.*, 1983). Negatively supercoiled DNA is preferentially assembled with histone octamers, because the association of each octamer with closed circular DNA will introduce a positive superhelical turn into the linker DNA (Section 2.2.3). It is more energetically favorable for octamers to bind to negatively superhelical DNA since the resultant positive supercoils in free DNA will be cancelled out by the existing negative supercoils (Clark and Felsenfeld, 1991). These correlations between chromatin assembly and transcription, which probably reflect the constraints of assembling any multicomponent protein complex on DNA, were next extended by examining the chromatin organization of specific DNA sequences containing well studied cellular genes.

Worcel and colleagues examined the organization into chromatin of the histone genes of *Drosophila* and *Xenopus* and the 5S RNA genes of *Xenopus*, after injection into oocyte nuclei as closed circular plasmid DNAs (Gargiulo and Worcel, 1983; Gargiulo *et al.*, 1985). A special technique was used to label closed circular DNA molecules with ^{32}P at a single restriction site (Fig. 3.1). This allowed the mapping of nuclease cuts on the assembled chromatin with great precision. It was found that changes in the amount of DNA injected into the oocyte nucleus influenced both the structure and expression of the assembled chromatin. Minichromosomes with a spaced array of nucleosomes, in which the distance between nucleosomes was identical to that found in the native chromosome, could be assembled under appropriate conditions. These parameters corresponded exactly to those for maximal transcription of the 5S RNA genes. Clearly tran-

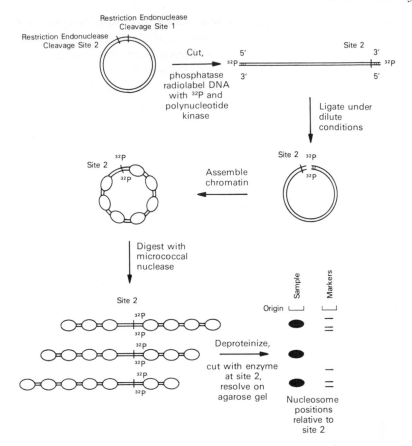

Figure 3.1.   Methodology for detecting specific chromatin structures on closed circular plasmid DNA molecules.
The radiolabeling is at site 1, mapping of nucleosome positions is relative to site 2.

scription of eukaryotic genes is not incompatible with the assembly of chromatin.

The correct separation between nucleosomes on small circular DNA molecules suggested that higher order chromatin structures such as the chromatin fiber were not required for the formation of correctly spaced arrays. This is because the plasmids used in these studies (3–5 kb in length) are too short to assemble more than 3–4 turns of a solenoidal chromatin fiber. Topological constraints make the formation of such a fiber unlikely. Surprisingly nucleosome positioning was not detected on the *Xenopus* 5S RNA or *Drosophila* histone gene in these experiments (Section 2.2.5). However, DNaseI hypersensitive sites did form 5' and 3' to the *Xenopus* histone gene, suggesting that under appropriate conditions specific chromatin structures could actually form on these artificial templates (Section 4.2.4). These hypersensitive sites depend on the association of specific non-histone proteins with the promoter.

The transcriptional activity of chromatin containing spaced nucleosomes should be contrasted with the inactivity of the same genes (5S RNA) assembled with nucleosome structures in which no linker DNA is present (Weisbrod *et al.*, 1982). Under these circumstances nucleosomes are described as close-packed. Thus incorrectly assembled chromatin structures are sometimes incompatible with transcription (Section 4.2.2).

Experiments examining the time-course of chromatin assembly under optimal conditions demonstrated that over 30 min were required to assemble DNA into a spaced array of nucleosomes in the oocyte nucleus; however, a 10.4 bp periodicity in DNaseI cleavage was apparent almost immediately (Ryoji and Worcel, 1984). This suggests that the wrapping of DNA around the core histones is a rapid event, but that the subsequent organization into a nucleosomal array is a slow process. As we will see later, many of these observations concerning chromatin assembly in the nucleus of the living oocyte can be reproduced *in vitro* (Section 3.4.2). The use of cloned DNA as templates for chromatin assembly has a great potential for correlating structure and function.

Unlike microinjection into oocyte cytoplasm, introduction of plasmid DNA molecules into the cytoplasm of *Xenopus* eggs led to their assembly in chromatin and stabilization against degradation by nucleases. However, detailed studies on the chromatin conformation of specific genes have not been performed. This major difference between the capacity of egg and oocyte cytoplasm to assemble chromatin is a consequence of the breakdown of the oocyte nucleus following the hormonally induced maturation of an ooocyte into an egg (Section 3.1.1). The contents of the oocyte nucleus remain in the egg cytoplasm where, following fertilization, they are used to assemble embryonic nuclei.

A major advance in our understanding of how chromosomes are assembled came from the utilization of the stores of nuclear components in the egg following injection of prokaryotic DNA into *Xenopus* eggs. DNA purified from bacteriophage λ was rapidly assembled into chromatin, surrounded by both a nuclear lamina and a nuclear envelope containing nuclear pores (Forbes *et al.*, 1983). These results demonstrated that nuclear assembly requires a DNA template and that specific DNA sequences are not necessary for this process to occur (Fig. 3.2). The spontaneously formed nuclei are not only struc-

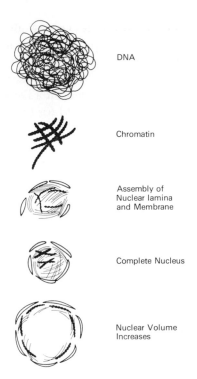

DNA

Chromatin

Assembly of
Nuclear lamina
and Membrane

Complete Nucleus

Nuclear Volume
Increases

Figure 3.2. Nuclear assembly visualized in *Xenopus* eggs or egg extracts.
The staged assembly of a nucleus around DNA is shown.

turally similar to endogenous nuclei, but also contain the molecular components necessary to respond to the cues that regulate nuclear structure during the cell cycle. For example, when the egg is arrested at mitosis, the bacteriophage λ DNA condenses into a thick filament resembling a metaphase chromosome. However, it should be noted that neither a centromere or a telomere form on this DNA, suggesting a requirement for specific DNA sequences in order to form these structures. Importantly the actual process of chromosome condensation does not require any specific DNA sequences in order to be regulated. These observations are consistent with the hypothesis that modulations in chromosomal structure are driven primarily through post-translational modification of the constituent chromosomal proteins (Sections 2.5.2 and 2.5.3).

Summary
Microinjection of exogenous DNA into *Xenopus* oocyte nuclei leads to

its assembly into nucleosomal arrays, some of which can be quite specific in their organization over promoter elements. Similar experiments in eggs lead to the formation of nuclei including the microinjected DNA even when derived from prokaryotic sources. This demonstrates that nuclei are assembled on a DNA template and that specific DNA sequences are not required for this assembly.

3.2.2 Chromatin assembly on DNA introduced into somatic cells

Although work with *Xenopus* eggs and oocytes provided our initial insights into how chromatin is assembled using exogenous DNA, obviously the general relevance and applicability of these results had to be established in somatic cells. There are several ways of introducing exogenous DNA into cells in culture or, more interestingly, into *Drosophila* or mammalian embryos. We have already discussed the first approach of fusing cells together to form somatic cell hybrids or heterokaryons. In recent times microinjection of *Drosophila* and mammalian eggs has yielded much important information concerning chromosomal structure and nuclear processes. However, these studies were pioneered by the transient or stable transformation of mammalian cells with plasmid DNAs or by infection with viral DNA.

The small size of the SV40 genome (5243 bp in length) and the early availability of information concerning DNA sequence and gene organization made it a convenient model for studying the structure and function of chromatin (Section 3.2.1). Late in infection the SV40 genome is organized into nucleosomal arrays as a minichromosome, which can be isolated from infected cells under appropriate conditions. One region of the minichromosome (ORI) contains several important recognition sites for *trans*-acting factors: the origin of replication, the binding sites for the viral regulatory protein (T-antigen) and the promoters of early and late SV40 mRNAs (Fig. 3.3). The chromosomal organization of this region was recognized as being important for the processes of DNA replication and transcription, and was shown to differ from the rest of the minichromosome.

The first experiments used nucleases to digest the minichromosome. The DNA sequence in the ORI region is preferentially cut in the nuclei of infected cells with DNaseI (a hypersensitive site, Section 4.2.4). This suggested that DNA in the ORI region is more accessible to DNA binding proteins than in the rest of the minichromosome. Chemical carcinogens such as psoralen that interact with the free DNA duplex by intercalation, bind preferentially to the ORI region in

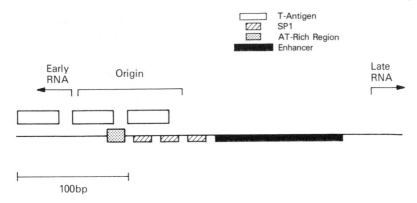

Figure 3.3. DNA sequence organization of the SV40 ORI region.
Binding sites for many *trans*-acting factors are compressed into this small segment of DNA. The origin and sites of transcription initiation are indicated, as are the major binding sites for *trans*-acting factors.

infected cells. Also consistent with a preferential accessibility of the ORI region is its rapid digestion with a variety of endonucleases in isolated minichromosomes. Finally, electron microscopy reveals that 20–25% of the isolated minichromosomes contain a nucleosome-free region (or gap) of approximately 350 bp covering the ORI region. This gap therefore represents the best early documentation of nucleosome positioning on DNA (Varshavsky *et al.*, 1978; Jacobovits *et al.*, 1980; Robinson and Hallick, 1982; Cereghini and Yaniv, 1984; Section 4.2.3).

An important question that is still to be resolved is whether this nucleosome-free region was imposed upon the chromosome by the interaction of *trans*-acting factors with their cognate DNA sequences, or whether histone–DNA interactions determine the placement of nucleosomes (sequence-directed positioning, Section 2.2.5). Recent studies have shown that while minor preferences exist in nucleosome positioning in the presence of only histones and DNA, other components are required to create a nucleosome-free gap (Weiss *et al.*, 1985; Ambrose *et al.*, 1989). We will see that the influence of non-histone proteins on nucleosome positioning is a common feature of chromatin structure that has important functional consequences (Section 4.2.5). In contrast to the failure of the ORI DNA sequence to exclude nucleosomes, similar reconstitutions of purified histones and DNA revealed a very favorable sequence for forming a nucleosome at the opposite end of the minichromosome. This is the region of the SV40 minichromosome where both replication and transcription terminate

(Poljak and Gralla, 1987; Hsieh and Griffith, 1988). It is also possible that a nucleosome at this position might assist the termination process by causing processive enzyme complexes to pause (Sections 4.3.1 and 4.3.4). These early experiments with SV40 minichromosomes provide strong support to the hypothesis that *trans*-acting factors function in an organized chromosomal environment.

Several genes or promoter elements have been introduced into SV40 minichromosomes and the influence of chromatin structure on gene expression examined. Although in general repressive effects on transcription are observed, the structural basis of this repression has not been determined (Lassar *et al.*, 1985). The advantage of viral genomes for the analysis of the interrelationship between chromatin structure and function is that they contain replication origins. This means that the viral minichromosome can be studied as an episome, i.e. without being integrated into the cellular chromosomes. Moreover, as chromatin assembly is coupled to DNA replication (Sections 3.3 and 3.4.3), the influence of *trans*-acting factors on nucleosome organization will be more like the true chromosomal context (Section 4.3.1). The only reservation about studies with small viral genomes is that possible regulatory effects dependent on higher order chromatin structure are unlikely to be observed. Again this is due to topological constraints in folding small DNA circles in the chromatin fiber. Viral genomes other than SV40 have been very useful for the detailed analysis of chromosomal influences on transcription. These include the bovine papilloma virus (BPV)-based episomes that have been used to investigate the molecular mechanisms by which glucocorticoid receptor activates gene expression (Ostrowski *et al.*, 1983) (Section 4.2.4). Here, the interplay between specific chromatin structures and transcription factors has been rigorously documented.

Viral episomes have been used to establish the organization of *trans*-acting factors and nucleosomes on specific DNA sequences; as we will see there is often interaction between these two components (Section 4.2.3). However, most scientists have investigated gene regulation by transiently transfecting cloned DNA without a viral origin of replication into eukaryotic cells. Under these conditions these investigators never encounter repressive or stimulatory effects that might be attributed to chromatin. Importantly, the observed regulation due to enhancers, promoters or other elements often does not reflect the range of response observed in the natural chromosomal context (Section 4.2.4). Early studies suggested that transfected plasmids that did not integrate into the natural chromosomal context would be assembled into nucleosomal arrays, even if they did not replicate in the cell (Camerini-Otero and Zasloff, 1980). However, more extensive studies by Howard and colleagues demonstrated that the efficiency of chromatin assembly depended upon the transfection conditions. The amount of DNA transfected into cells, the method of compacting DNA prior to transfection and the efficiency with which particular DNA sequences partitioned to the nucleus greatly affected chromatin assembly. Over 80–90% of nuclear plasmid material might not be assembled into chromatin (Reeves *et al.*, 1985). Consequently it is not surprising that the regulation of transiently transfected DNA does not completely reflect that found in the natural chromosomal context.

It is possible to select for the stable chromosomal integration of transfected DNA in somatic cells using genetic markers conferring resistance against certain drugs. These studies have revealed that gene expression varies depending on the site of integration within the chromosome. This is another manifestation of a position effect (Section 2.5.5). For example, an intact single copy of the immuno-globulin K gene containing 1.5 kb of upstream and 8.5 kb of down-stream flanking sequences exhibited a 100-fold variation in transcriptional activity dependent on chromosomal position. This probably reflects the proximity of the integrated DNA sequence to heterochromatin (Section 2.5.5) or other repressive influences such as silencers (Section 4.1.2). Expression of the stably integrated DNA was found to be only a quarter of that expected for the endogenous gene when a large number of transformed cell lines were examined. Clearly a significant factor required for efficient gene expression lies in having the correct chromosomal organization (Blasquez *et al.*, 1989).

The correct chromatin organization for gene expression has only been determined for a few genes (Section 4.2.3). Although the position of individual nucleosomes has generally not been examined in this type of experiment, investigators have searched for sequence specific influences on the organization of nucleosomal arrays and the chromatin fiber. Specific DNA elements distant from the gene itself appear to confer insulation from chromosomal position effects (Section 4.2.4). At a more fundamental level, removal of all non-coding sequences other than the promoter from the immunoglobulin K gene in stable integrants, leads to a reduction of transcriptional activity to only 2% that of wild type. These experiments introduce the role of elements other than the promoter that act at a distance to influence transcription initiation (Section 4.1.2) and that may also influence chromosomal organization (Section 4.2.4).

The method of choice for introducing DNA into mammalian chromosomes, such that it will be both expressed and regulated in the correct way, is to microinject it into mammalian eggs. Similar experiments have been carried out in *Drosophila*. This approach allows both the stable integration of the microinjected DNA into the chromosome and facilitates the analysis of the function of various *cis*-acting elements in regulating genes *in vivo* in the normal developmental environment (Grosveld *et al.*, 1987). An additional advantage is the capacity to assess the tissue-specific expression of genes integrated at identical chromosomal positions in each of the different cell types present in the animal. These studies have facilitated the definition of regulatory elements called locus control regions (LCRs). These DNA sequences apparently shield the gene from the influence of chromosomal position, i.e. position effects do not occur. The evidence to support this hypothesis is discussed later (Section 4.2.4) but in general the presence of LCRs allows position-independent, copy-number-dependent expression of a gene. Similar elements have also been defined in *Drosophila*, where once again the elements are believed to define a functional domain of chromatin (Kellum and Schedl, 1991).

Summary
The assembly of exogenous DNA into chromatin in somatic cells depends both on the organization of the DNA and how it is introduced. Use of constructs containing origins of replication (e.g. viral DNAs) leads to chromatin structures that resemble those found in the chromosome itself. Microinjection of DNA into fertilized eggs (transgenic experiments) has led to the definition of DNA elements known as locus control regions that may organize whole domains of chromatin.

3.2.3 Yeast minichromosomes

A particularly attractive system for examining the influence of chromatin structure on the function of DNA is yeast. Experimental work with yeast has many advantages, especially for molecular biologists and geneticists; among these is the existence of small ($<$ 1500 bp) extrachromosomal plasmids that will replicate autonomously. Particular experimental attention has been given to a plasmid present at about 100 copies/cell known as TRP1ARS1 (1453 bp in length). This plasmid consists of one gene coding for N-(5'-phosphoribosyl)anthranilate isomerase (*TRP1*), a sequence containing a replication origin (ARS1), and a segment of unknown function (UNF) (Fig. 3.4). Simpson, Thoma and colleagues have determined the chromatin structure

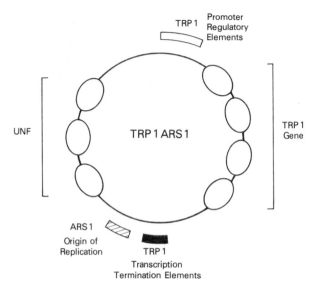

Figure 3.4. The structure of the TRP1ARS1 minichromosome. Nucleosomes (open ellipsoids) and key DNA sequences are shown.

of TRP1ARS1 in some detail (Thoma and Simpson, 1985; Simpson, 1991). Nucleosome position has been determined by the nuclease accessibility of isolated minichromosomes. Four different regions of chromatin structure have been defined. The UNF region contains three nucleosomes that have very strong DNA sequence-directed positions (Section 2.2.5), which are not changed by insertion of DNA fragments of various lengths into the plasmid. In contrast, four loosely positioned nucleosomes are found on the *TRP1* gene. These nucleosomes can easily rearrange following insertion of additional DNA. Two nucleosome-free regions are also present (like the gap in the SV40 minichromosome; Section 3.1.4) which are hypersensitive to nuclease digestion. These sequences include the TRP1 promoter and the ARS origin of replication. Although lacking centromeres and telomeres the TRP1ARS1 minichromosome therefore contains many elements associated with normal cellular chromosomes.

Among the DNA sequences inserted into the TRP1ARS1 plasmid was a sea urchin 5S RNA gene, on which base pair resolution of the rotational positioning of DNA on the histone core had originally been defined (Section 2.2.5). A nucleosome formed including this sequence, with exactly the same position in the yeast minichromosome as observed *in vitro*. This important observation shows that yeast histones *in vivo* are able to recognize the same DNA sequence-directed nucleosome positioning elements as chicken histones *in vitro*

(Thoma and Simpson, 1985). Similar experiments with yeast genomic sequences inserted into the TRP1ARS1 plasmid consistently reveal the same nucleosome positioning to occur on the episome as seen in the chromosome (Simpson, 1991). In certain instances unstable nucleosomes such as those on the *TRP1* gene may have their positions influenced by the organization of other regions of the episome into chromatin (Thoma and Zatchej, 1988). More recent studies have taken these observations further to dissect the contributions of specific *trans*-acting factors to nucleosome positioning.

A segment of DNA containing the binding site for a *trans*-acting factor (the α_2-repressor) was inserted into the TRP1ARS1 plasmid. The α_2-repressor is a protein produced by α-mating type cells that binds to DNA as a complex with a second protein, MCM1, to repress transcription of genes normally expressed only in an α-mating type cell. When α_2 was not present the chromatin structure of the inserted DNA segment appeared to be random; no nucleosome positioning was apparent on micrococcal nuclease digestion. However, the presence of α_2 led to a dramatic change in chromatin structure (Roth *et al.*, 1990). The entire minichromosome, except for the nucleosome-free region around the origin of replication, was organized into an array of precisely positioned nucleosomes. A direct interaction between the α_2-repressor and the nucleosome has been suggested by the capacity of the protein to move a nucleosome adjacent to it apparently without any influence from the underlying DNA sequence. Moreover, protein–protein contacts between the α_2-repressor and histones are suggested by the observation that two yeast mutants in which the N-terminal tail of histone H4 is deleted, do not form positioned nucleosomes in the presence of the α_2-repressor (Simpson, 1991; Section 4.2.5). These studies have been extended to show that the association of the α_2-repressor with its binding site will cause a nucleosome to be positioned over the TATA box of yeast chromosomal α-mating type specific genes *in vivo* (Shimizu *et al.*, 1991). The demonstration of specific interactions between non-histone proteins and histones that contribute to nucleosome positioning introduces a new level of complexity and specificity to the assembly of chromatin and chromosomes (Section 4.2.5).

Summary

Yeast minichromosomes have provided major insights into nucleosome positioning. DNA sequence- and *trans*-acting factor-directed nucleosome positioning have been defined. It has been established that specific interactions exist between the histone proteins and *trans-*acting factors responsible for directing nucleosome position.

3.3 CHROMATIN ASSEMBLY ON REPLICATING ENDOGENOUS CHROMOSOMAL DNA *IN VIVO*

A large fraction of histone synthesis in somatic cells (unlike *Xenopus* oocytes and eggs) is coupled to DNA replication during the S-phase of the cell cycle. As the DNA content of the chromosome is doubled, so must the protein component be duplicated to reconstruct two daughter chromosomes. Several experiments in which newly synthesized chromatin (nascent chromatin) was fractionated from pre-existing chromatin (old chromatin) have shown that chromatin assembly *in vivo* is a staged process. Worcel exploited the increased susceptibility of nascent chromatin to nucleases to fractionate nascent chromatin from *Drosophila* embryos. Newly synthesized DNA was enriched for newly synthesized histones H3 and H4, while newly synthesized histones H2A/H2B and H1 associate with chromatin that had properties similar to those of bulk non-replicating chromatin. Worcel concluded that newly synthesized histones associated with newly synthesized DNA in a sequential order: histones H3 and H4 are deposited first, then histones H2A and H2B, and finally histone H1 (Worcel *et al.*, 1978; Fig. 3.5).

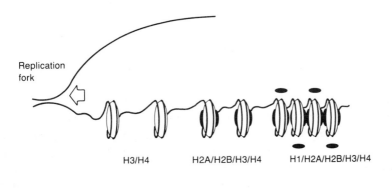

Figure 3.5.   Chromatin assembly on replicating DNA.
Maturation of chromatin is shown from left to right. Binding of $(H3/H4)_2$ tetramers to nascent DNA is followed by H2A/H2B dimers and finally histone H1.

Experiments with mammalian cells have reached remarkably similar conclusions with respect to the staged assembly of chromatin *in vivo*. The process of chromatin maturation on newly synthesized DNA takes several minutes even in a rapidly proliferating mammalian cell. Nascent DNA is assembled into a structure that is more sensitive to nucleases than mature chromatin. The loss of this nuclease-sensitive conformation takes over 10–20 min (Cusick *et al.*, 1983). Chalkley and co-workers have suggested that the initial rapid deposition of histones H3 and H4 on newly synthesized DNA reflects the nuclease-sensitive stage, whereas the subsequent deposition of histones H2A and H2B correlates with the appearance of regular nucleosomal arrays and nuclease resistance (Smith *et al.*, 1984). The sequential sequestration of histones is clearly related to the structure of the nucleosome, since histones H3 and H4 form the core of the structure, whereas histones H2A and H2B bind at the periphery of the nucleosome and histone H1 can only associate in its proper place once two turns of DNA are wrapped around the core histones (Sections 2.2.2 and 2.2.4). We will see later that this two-stage maturation process has important implications for the accessibility of *trans*-acting factors to newly synthesized DNA (Section 4.3.1).

Newly synthesized histones that are destined for deposition on nascent DNA possess transient post-translational modifications that distinguish them from the 'old histones' of bulk chromatin. One modification studied in some detail is the diacetylation of histone H4 at specific lysine residues. This is true not only for newly synthesized H4 in mammalian cells, but is also the form in which histone H4 is stored in *Xenopus* oocytes, the form present in cleavage stage embryonic chromatin in the sea urchin and in replicating nuclei in *Tetrahymena* (Csordas, 1990). Approximately 30–60 min after deposition during chromatin assembly, the diacetylated histone H4 is deacetylated to its mature form. Histone H3 is also acetylated, but is deacetylated more rapidly (1–2 min) than histone H4 in nascent chromatin. Acetylation may actually facilitate deposition of histones H3 and H4 at the replication fork (Section 3.4.3). Moreover, acetylation of the histone tails may partially account for the lack of compaction of nascent chromatin, the reduction in histone H1 content, the accessibility of this chromatin to nucleases and perhaps to *trans*-acting factors (Section 2.5.2).

Annunziato and colleagues have examined the significance of acetylation for nascent chromatin using the deacetylase inhibitor sodium butyrate. This inhibitor has no effect on the DNA replication process; however, newly synthesized histone H4 retains its nascent, diacetylated form. Normally, nascent chromatin would mature to a nuclease-resistant form after 10–20 min. In the presence of butyrate nascent chromatin never achieves the nuclease resistance of bulk chromatin. The nascent chromatin does, however, assemble correctly spaced nucleosomes, demonstrating that acetylated histones can be assembled into chromatin resembling that found in a normal chromosome. Subsequent experiments have shown the observed difference in nuclease sensitivity between nascent and bulk chromatin to be related to the binding of histone H1 (Perry and Annunziato, 1989). Histone H4 deacetylation appears to be required for this linker histone to bind. Consistent with other observations (Section 3.4.2) these results demonstrate that histone H1 is not required for the assembly of spaced nucleosomal arrays.

Histone H1 itself is either unmodified or moderately phosphorylated in nascent chromatin. As chromatin matures, histone H1 becomes more highly phosphorylated (Jackson *et al.*, 1976). As we have discussed (Section 2.5.3), the effect of phosphorylating histone H1 is most likely to weaken its binding to linker DNA by altering the level of charge neutralization of linker DNA. Consequently we might expect the dephosphorylated histone H1 to exhibit a tighter interaction with DNA even though the highly acetylated state of histone H4 might prevent correct interaction with the core histones.

Not all chromatin assembly in the cell has to occur during replication. Normally the vast majority of histone synthesis is coupled to DNA synthesis. However, Bonner, Zweidler and colleagues have identified specific variants of histones H2A and H2B encoded by distinct genes that are synthesized in G1 and G2 as well as in S-phase. This constitutes a basal level of histone synthesis which varies from organism to organism. In mammalian cells it can constitute approximately 4% of total histone production. The histones synthesized in G1 enter chromatin in the absence of replication, demonstrating an active and efficient exchange process with pre-existing histones (Bonner *et al.*, 1988; Zweidler, 1980; Section 3.1.2). Chalkley and colleagues have shown that histones H1, H2A and H2B synthesized in the presence of inhibitors of DNA synthesis will exchange with pre-existing histones in nucleosomes whereas histone H3 and H4 do not (Louters and Chalkley, 1985). It has been suggested that this exchange is facilitated by transcription (Jackson, 1990).

Summary

Chromatin assembly *in vivo* is coupled to replication. It is possible to divide this process into two stages: the initial deposition of histones H3/H4 followed by the sequestration of histones H2A/H2B and H1.

Histones H3/H4 are acetylated when they are initially incorporated into chromatin, and their progressive deacetylation correlates with the sequestration of histone H1 and chromatin maturation.

3.4 CHROMATIN ASSEMBLY *IN VITRO*

3.4.1 Purified systems

Once it was realized that high salt concentrations ($>$ 1.2 M NaCl) would dissociate nucleosome core particles into their components of DNA and histones, it was soon established that the process was reversible. This reversibility has been extremely useful in establishing the organization of both DNA and histones in the nucleosome (Section 2.2.5). Separation of H3/H4 tetramers from H2A/H2B dimers has allowed the demonstration that histones H3/H4 recognize DNA sequence-directed nucleosome positioning elements even when they first associate with DNA at high ionic strength ($\sim$ 1 M NaCl) (Hansen *et al.*, 1991). Histones H2A/H2B bring more DNA into the nucleosome when they bind ($<$ 0.8 M NaCl) however they do not alter the final position of the histone core relative to the double helix.

Although useful for examining the organization of DNA with a single histone octamer or tetramer, reconstitution of chromatin from high salt has several disadvantages. The most notable failing of this methodology is the inability of nucleosomes assembled in this way to space themselves correctly. Instead of the physiological spacing found in the chromosome, histone octamers reconstituted by dialysis from high salt concentrations pack together on DNA as closely as possible. This does not mean that all nucleosomes form adjacent to each other, only that when they are adjacent there is no linker DNA. In general one nucleosome is found every 150 base pairs whereas *in vivo* nucleosomes are spaced every 180–190 base pairs (depending on tissue and organism). This close packing appears to be a consequence of maximizing not only strong protein–DNA interactions, but also protein–protein interaction along the DNA backbone. *In vivo* nucleosomes are assembled in stages, the histones are extensively acetylated and thus have relatively weak protein–DNA contacts (Section 3.3). In contrast, following dialysis from high salt, an entire unmodified histone octamer is deposited onto DNA. Since there is very little linker DNA between these close packed nucleosome structures, histone H1 cannot bind correctly. Consequently histone H1 interacts with long, exposed stretches of free DNA forming aggregates or causing DNA to precipitate. Nevertheless, under highly controlled conditions, Stein and colleagues have been able to show that linker histones can influence the spacing of histone octamers, at least on the highly flexible oligo(dA.dT) DNA molecules (Stein and Bina, 1984; Section 2.1.1). These experiments encourage the hope that methodologies can eventually be found that lead to the appropriate association of histone H1 with a synthetic chromatin substrate.

Simpson and colleagues took a novel approach to this question by constructing a template that avoids all the problems we have just discussed and assembles correctly spaced chromatin following dialysis of histone and DNA from high to low salt concentrations. This difficult task was accomplished by spacing the nucleosome-positioning elements in the sea urchin 5S RNA gene in tandem arrays, with repeat lengths spanning the range of most cellular chromatins. The long, tandemly repeated (> 50) DNA fragments were then excised and reconstituted with core histones. The DNA sequence-directed nucleosome positioning signals were of sufficient strength to overcome the tendency of histone octamers to close pack (Simpson *et al.*, 1985). We have already discussed the utility of these spaced nucleosomal arrays in examining their salt dependent compaction into the chromatin fiber (Section 2.3.1). More recent studies have demonstrated that even histone H1 might bind correctly to these arrays (Meersseman *et al.*, 1991). Bradbury and colleagues made use of an analysis of minor translational positions of histone octamers on the sea urchin 5S DNA to show that histone H1 could influence translational position. Histone H1 was mixed with the chromatin at high salt concentrations (0.5 M NaCl) and dialyzed to low ionic strength, leading to its association with chromatin without causing precipitation. This system offers considerable hope for the correct reconstitution of histone H1 into chromatin.

Obviously the mechanism of chromatin assembly *in vivo* differs considerably from the *in vitro* dialysis of mixtures of histone and DNA from high to low salt. However, there are several examples of macromolecules that perhaps resemble salt in their effects and interaction with the core histones. These acidic macromolecules such as polyglutamic acid have been used to assemble chromatin. In many ways they resemble physiological nucleosome assembly factors (Section 3.4.2). At physiological ionic strengths (0.2 M NaCl) nucleosome assembly normally occurs very inefficiently; however, it can be accomplished by renaturing core histones into H3/H4 tetramers and H2A/H2B into dimers by prolonged dialysis. These proteins are then titrated over another protracted period with a considerable excess of naked DNA. The addition of limiting histones prevents the formation of large aggregates and the precipitation of the histone–DNA complex (Ruiz-Carrillo *et al.*, 1979). Obviously this process needs to be considerably improved *in vivo*. In order to understand chromatin assembly *in vivo* we have to introduce components other than histones and DNA. These represent a special class of macromolecules known as molecular chaperones.

Summary
It is possible to assemble nucleosomal structures by the progressive dialysis of mixtures of DNA and histones from high to low salt. Unless special nucleosome positioning sequences are used the nucleosomes will pack together very closely, unlike the correct physiological spacing seen in the chromosome. Constructs that do position nucleosomes may allow histone H1 to associate in the correct way, thereby assembling completely synthetic chromatin.

3.4.2 Nucleosome assembly in extracts of *Xenopus* eggs and oocytes

Major advances in understanding chromatin assembly processes *in vitro* followed from the discovery that the *Xenopus* egg contained large stores of histones which could provide a source of chromatin assembly materials. It was possible to prepare extracts of *Xenopus* eggs, mix in purified homogeneous DNA and assay for chromatin assembly. The definition of the chromatin structure of SV40 minichromosome provided a reference product for assaying nucleosome assembly on a small closed circular DNA molecule. Laskey and colleagues prepared homogenates of *Xenopus* eggs in which SV40 DNA would rapidly (less than 1 h) be assembled into a minichromosome under apparently physiological conditions. The assembly of chromatin was found not to be cooperative, nor did it require synthetic processes such as DNA replication or protein synthesis. Biochemical fractionation of the extract led to the recognition that at least three protein components were required: histones, topoisomerases and nucleoplasmin (Laskey and Earnshaw, 1980).

Nucleoplasmin is an acidic thermostable protein which promotes nucleosome assembly from purified histones and DNA, but only when present in excess over histones. The interaction of nucleoplasmin with histones in the absence of DNA led to the use of the term chaperone to describe its properties; in effect, nucleoplasmin prevented the uncontrolled association of histones with the double helix.

Nucleoplasmin is the most abundant protein in the nuclei of *Xenopus* oocytes representing up to 10% of the total nuclear protein (Mills *et al.*, 1980). Nucleoplasmin is rich in acidic amino acids (~ 20% aspartic and glutamic acid), and is also heavily phosphorylated, making the protein even more acidic. The active form of nucleoplasmin for nucleosome assembly is a pentamer of 22,024 Da subunits. Interestingly, nucleoplasmin does not appear to interact directly with DNA or chromatin, but will bind histones *in vitro*. Immunolocalization shows that the protein is associated with actively transcribed DNA, possibly reflecting the dynamic state of nucleosomes during transcription (Moreau *et al.*, 1986; Section 4.3.4). Nucleoplasmin has been cloned and sequenced revealing that the acidic amino acids form several clusters. One region contains 17 acidic amino acids (15 glutamic acid and 2 aspartic acid) out of 20 in the middle of the monomer. The acidic character of nucleoplasmin led Stein and colleagues to demonstrate that polyglutamic and polyaspartic acid can greatly facilitate nucleosome assembly at physiological ionic strength. Moreover, Stein showed that polyglutamic acid could actually cause histones to associate as stable octamers at physiological ionic strength in the absence of DNA. These observations make the polyacidic tracts obvious candidates for the histone binding site in nucleoplasmin (Stein *et al.*, 1979).

Although it was originally thought that nucleoplasmin was the only molecular chaperone required for nucleosome assembly, more recent results have indicated a more specific role. Kleinschmidt and colleagues identified a second protein within the nuclei of *Xenopus* oocytes called N1/N2, that forms a specific complex with only histones H3 and H4. N1/N2 was cloned and found to be a 64,774 Da protein, also containing polyacidic tracts. One tract contains 18 glutamic acid and 3 aspartic acid residues out of a run of 31 amino acids. This particular site has been shown by mutagenesis to be important for the binding of histones H3/H4 to N1/N2 (Kleinschmidt *et al.*, 1985, Kleinschmidt and Seiter, 1988).

The preparation of specific antibodies to both nucleoplasmin and N1/N2 allowed the resolution of their respective roles in nucleosome assembly in *Xenopus* egg extracts. Histones H3/H4 are found to co-immunoprecipitate with N1/N2, whereas histones H2A/H2B are found to co-immunoprecipitate with nucleoplasmin. Subsequent work has shown that each complex transfers histones to DNA separately. Histones H3/H4 have to be associated with DNA before nucleoplasmin can deposit histones H2A/H2B (Dilworth *et al.*, 1987; Kleinschmidt *et al.*, 1990). This ordered sequestration of histones is similar to that occurring at the replication fork *in vivo* (Section 3.3), although different proteins appear to be involved in somatic cells (Section 3.4.3). An important point from these observations is that although many polyanions such as polyglutamic acid, proteins containing polyacidic tracts (like HMG1 or CENP-B) or even ribonucleic acid can assemble nucleosomes, this assembly process may not be physiologically relevant (Nelson *et al.*, 1981; Bonne-Andrea *et al.*, 1984). Indeed although nucleoplasmin and N1/N2 are now well established as nucleosome assembly proteins in *Xenopus* eggs, it is unclear whether they fulfill comparable roles in somatic cells.

The original cell-free preparation of *Xenopus* eggs capable of assembling chromatin involved homogenizing eggs before insoluble yolk was removed by gentle centrifugation. Improvements to this protocol came from the observation of Lohka and Masui that gentle centrifugation of eggs would lead to the stratification of the egg contents (Lohka and Masui, 1983; Fig. 3.6). Following this it was relatively easy to

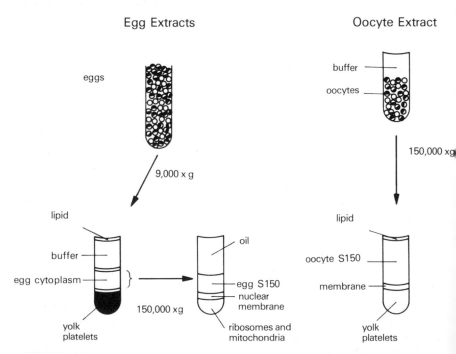

Figure 3.6.   Preparation of chromatin assembly extracts from *Xenopus* eggs and oocytes.
A single centrifugation step (oocyte) or two centrifugation steps (egg) are shown.

isolate a fraction greatly enriched in egg cytoplasm. An important advantage of this protocol was that it enabled egg cytoplasmic preparations to be made that were essentially free of yolk. Yolk contains highly phosphorylated polypeptides that exert a strong inhibitory effect on any process involving nucleic acid (Wolffe and Schild, 1991). Worcel and colleagues adapted this protocol for *Xenopus* oocytes; the high-speed supernatant of oocytes (after centrifugation at 150,000 × *g*, known as the oocyte S150) has been used in a number of highly significant experiments.

An important aspect of the *Xenopus* chromatin assembly systems is that the resultant nucleosomal arrays are physiologically spaced (i.e. one nucleosome every 180–190 bp) identical to their spacing in *Xenopus* chromosomes. As we have discussed, the failure to assemble spaced nucleosomal arrays has presented an important limitation to chromatin assembly using purified components (Section 3.4.1). The *Xenopus* oocyte S150 system requires Mg^{2+} and ATP to assemble these correctly spaced nucleosomal arrays. Although the exact spacing of nucleosomes is influenced by the excess of histones over DNA in the extract, close-packed arrays of nucleosomes are difficult to form.

The assembly of spaced nucleosomes initially presented an enigma to investigators as histone H1 is not present in the minichromosomes assembled *in vitro*. Previous work using synthetic DNA templates (oligo(dA.dT)) and purified histones had suggested that histone H1 was involved in determining nucleosome spacing (Section 3.4.1). Consistent with the absence of histone H1 from the oocyte S150 assembly system, but also in keeping with the results using purified components, Worcel and colleagues were able to show that exogenous histone H1 could influence the spacing of nucleosomes on DNA. Nucleosomal repeats as long as 220 bp could be obtained (Rodriguez-Campos *et al.*, 1989). Worcel also studied the possible role of histone acetylation in the spacing of nucleosomes. Although histone H4 was deacetylated during the maturation of chromatin in the *Xenopus* oocyte S150, this deacetylation process could be completely inhibited without influencing nucleosome spacing (Shimamura *et al.*, 1989). Thus these experiments with oocyte extracts eliminate two major candidates for spacing nucleosomes: histone H1 and the acetylation of histone H4. However, whatever is responsible for spacing nucleosome requires exogenous Mg^{2+}/ATP *in vitro*.

Similar experiments have been performed using similar preparations derived from *Xenopus* eggs. Eggs have several advantages over oocytes for the preparation of chromatin assembly extracts. The egg is a well-defined biochemical entity, whereas an oocyte may be within one of several developmental stages each containing a distinct store of macromolecules (Section 3.1.1). Aside from the physiological integrity of the egg compared to an oocyte, the molecules involved in chromatin assembly may be more active when isolated from eggs. There is evidence that this is the case for *Xenopus* nucleoplasmin. The more highly phosphorylated form of the protein is more competent for depositing histones onto DNA, this form is more abundant when isolated from eggs rather than oocytes (Sealy *et al.*, 1986). Like the oocyte S150, the egg extract requires supplementation of the *in vitro* reaction with exogenous Mg^{2+}/ATP in order to assemble spaced chromatin. Almouzni and colleagues showed that this assembly of spaced nucleosomes did not require ATP hydrolysis, suggesting instead that perhaps the stabilization of the phosphorylation state of an important component of the system might be important. Mg^{2+}/ATP acts as a competitive inhibitor to prevent the action of phosphatases. This idea receives further support from the observation that the efficient deposition of histones H2A and H2B into chromatin required exogenous Mg^{2+}/ATP. Phosphorylation of nucleoplasmin might be expected to promote this process (Almouzni and Mechali, 1988a; Almouzni *et al.*, 1991). Furthermore, it has recently been proposed that the phosphorylation of a variant of histone H2A might be important in mediating the correct spacing of nucleosomes (Kleinschmidt and Steinbeisser, 1991). Thus the assembly of spaced nucleosomal arrays is likely to be a property of the core histones dependent upon their post-translational modification or that of a molecular chaperone.

The egg extract has the capacity to carry out second strand synthesis of a single-stranded DNA molecule. This replication process resembles the enzymatic activities associated with lagging strand synthesis at the replication fork in the *Xenopus* embryo. Most importantly the synthesis of duplex DNA is very efficient and rapid, occurring at comparable rates to those observed *in vivo*. Using this system, Almouzni and colleagues were able to show that nucleosome assembly was much more rapid on the replicating template than on non-replicating duplex DNA mixed with the extract. Dissection of the molecular basis of this effect revealed that chromatin assembly was staged on the replicating template such that histones H3 and H4 were rapidly deposited on DNA to form an intermediate complex. Following assembly of histone H3/H4 tetramers, histones H2A/H2B were also preferentially sequestered on to the replicated DNA (Almouzni and Mechali, 1988b; Almouzni *et al.*, 1990a, b). This preferential association of histones with the replicating template was not due to a selective affinity for single-stranded DNA, instead it appears coupled to the replication machinery itself. This observation is important because *in vivo* chromatin assembly is coupled to replication (Section 3.3.). Thus, two systems from *Xenopus* can assemble spaced arrays of nucleosomes on exogenous DNA in vitro. Later we will discuss how nuclei are assembled using this type of chromatin (Section 3.4.4).

Summary

Biochemical fractionation of *Xenopus* egg extracts led to the discovery of molecular chaperones: nucleoplasmin and N1/N2. Both are acidic proteins that interact with histones to form specific complexes. Nucleoplasmin binds histones H2A/H2B and N1/N2 binds histones H3/H4. N1/N2 has to deposit histones H3/H4 onto DNA before H2A/H2B can be sequestered.

Whole-egg and oocyte extracts assemble correctly spaced nucleosomal arrays. Histone H1 and the acetylation of histone H4 are not involved in the spacing phenomenon. Histones H2A/H2B, especially phosphorylated histone H2A, appear to be important in determining the physiological spacing of nucleosomes. DNA replication in whole egg extracts has been shown to facilitate nucleosome assembly, once again in a staged process.

3.4.3 Nucleosome assembly in mammalian cell extracts

Nucleosome assembly systems derived from mammalian cells have been developed. These experiments are based on an *in vitro* replication system developed by Kelly and colleagues that makes use of the SV40 origin of replication and a viral protein, T antigen (Li and Kelly, 1984). Whole-cell extracts have been studied that are dependent on DNA replication for efficient nucleosome assembly. Rather less work has been carried out on the replication-independent system. Nucleosome assembly in mammalian whole cell extracts is also promoted by the replication process. Like the *Xenopus* egg and oocyte extracts, the assembly of regular arrays of nucleosomes does not require histone H1, but does require exogenous Mg^{2+}/ATP.

As discussed earlier, SV40 replicates in the nucleus of the host cell as a circular chromosome whose nucleosome structure and histone composition are identical with those of the host (Section 3.2.2). With the exception of T-antigen, the only viral protein required for replication, all of the enzymology of replication and the assembly of the minichromosome are identical to normal cellular processes. The original extracts used for replication contained predominantly cytoplasmic components. Stillman and colleagues observed that the addition of an extract of mammalian cell nuclei promoted the assembly of nucleosomes on the replicating SV40 DNA. This nucleosome assembly was dependent upon and occurs concomitantly with DNA replication. Subsequent experiments fractionated the nuclear extract and showed that it contained a single component required for the replication-dependent nucleosome assembly (Smith and Stillman, 1989, 1991a, b). This nuclear protein, called chromatin assembly factor I (CAF-1) was purified to homogeneity and found to consist of five polypeptides of 150, 62, 60, 58 and 50 kDa. CAF-1 does not appear related to either nucleoplasmin or N1/N2, although some of the polypeptides are phosphorylated. Like chromatin assembly in *Xenopus* extracts, nucleosome assembly on the replicating SV40 template is a stepwise process. During the first step, CAF-1 targets the deposition of newly synthesized histones H3 and H4 to the replicating DNA. This reaction is dependent on and coupled to DNA replication. In the second step, the histone H3/H4 complex is converted to a mature nucleosome by the sequestration of histones H2A/H2B. This latter process can occur after replication is complete (Gruss *et al.*, 1990).

Summary

Mammalian whole-cell extracts are capable of replicating duplex DNA containing a viral origin of replication in the presence of a single viral replication protein. In this system nucleosome assembly is coupled to replication. Sequestration of both histones H3/H4 and histones H2A/H2B is strongly favored on replicating DNA.

3.5 NUCLEAR ASSEMBLY *IN VITRO*

We have discussed the assembly of nucleosomal arrays in *Xenopus* egg extracts (Section 3.4.2); this is, however, only the first stage in a much more complex process of assembling nuclei *in vitro*. The first successful approach to this problem was a direct extension of Gurdon's earlier nuclear transplantation experiments to the *in vitro* preparation of egg cytoplasm. Lohka, Masui and colleagues found that *Xenopus* sperm nuclei incubated in egg extracts decondensed and initiated DNA replication (Lohka and Masui, 1983, 1984; Fig. 3.7). Subsequent work has defined the decondensation process and subsequent assembly of 'normal nuclear architecture' as a multi-stage process. *Xenopus laevis* sperm nuclei contain normal histones H3 and H4, but have their H2A, H2B and H1 replaced by sperm-specific

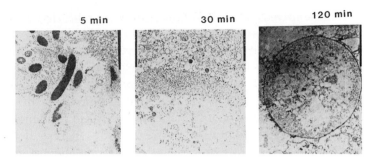

Figure 3.7. Remodeling of *Xenopus* sperm nuclei during incubation in the *Xenopus* egg extract.
Electron micrographs of sperm nuclei incubated in the extract for the times incubated are shown.

proteins (Poccia, 1986; Wolffe, 1989a, 1989b). These sperm-specific proteins dissociate from the sperm chromatin in the *Xenopus* egg extract (Wolffe, 1989b; Philpott *et al.*, 1991). This dissociation may be facilitated by the phosphorylation of these small basic proteins. It also appears that nucleoplasmin may facilitate both the removal of the sperm-specific proteins and the deposition of histones H2A/H2B to assemble normal nucleosomes (except for the absence of histone H1). This represents the first stage of decondensation of sperm chromatin, the second stage requires the assembly of a nuclear membrane on the surface of the remodeled chromosomes. Once a nuclear membrane is assembled, the nucleus swells to the normal volume expected for a nucleus within the *Xenopus* early embryo (see Fig. 3.2). What determines this volume is unknown.

The remodeling of sperm nuclei in *Xenopus* egg extracts strongly resembles events that occur in the living egg (Section 3.2.1). However, unlike the *in vivo* case, *in vitro* it is possible to dissect the process in some detail. Methodologies have been pioneered by examining the reorganization of specialized nuclei in *Xenopus* egg extracts. An important result is that it is also possible to assemble nuclei *de novo* on naked DNA. Newport and colleagues have followed this process in some detail. Building upon the observation that the injection of bacteriophage λ DNA into *Xenopus* eggs resulted in the efficient assembly of nuclei (Section 3.2.1), the similar process was followed *in vitro*. Immediately after addition to the extract, the λ DNA appeared as long, decondensed strands. After 20–30 min, the DNA was condensed into structures resembling the chromatin (30 nm) fiber and DNA became even more condensed over the following 40 min. At this time a nuclear envelope appeared around the chromatin and nuclear volume increased 20–30-fold concomitant with decondensation of the chromatin (see Fig. 3.2).

Topoisomerase II inhibitors were found to prevent nuclear formation by preventing the formation of highly condensed chromatin. More recently, Laemmli and colleagues have specifically depleted topoisomerase II from *Xenopus* extracts using antibodies. In these depleted extracts the capacity of the chromatin fiber to condense was severely inhibited. Re-addition of purified topoisomerase II (from yeast) to the depleted extract restored its capacity to convert decondensed chromosome to their mitotic state. Titration of the amount of topoisomerase II revealed that the high topoisomerase II concentration of mitotic extracts needs to be matched with the yeast enzyme in order to rescue the assembly potential of extracts depleted for this enzyme. One topoisomerase II protein per 2 kb of DNA leads to partial condensation, while three times as much leads to complete condensation of chromosomes (Luke and Bogenhagen, 1989; Adachi *et al.*, 1991). These experiments dramatically confirm the central role of topoisomerase II in mediating the DNA rearrangements necessary for chromosome folding.

The next experiments examined the assembly of the nuclear lamina and membrane. *Xenopus* embryonic nuclei appear to only contain a single lamin species, L_{111}. Presumably this simplifies the structure and properties of the lamina. DNA at the highly condensed stage of nuclear reconstitution does not initially have detectable amounts of lamin associated with it. However, eventually a lamina and nuclear envelope form around the condensed chromatin and appear to promote the subsequent swelling of the nuclear structure. Assembly of the nuclear membrane appears to be an essential prerequisite for subsequent nuclear events such as the initiation of DNA replication (Leno and Laskey, 1991). Both the nuclear lamina and topoisomerase II are presumed to have key roles in chromosomal architecture as components of the nuclear scaffold (Section 2.4.2).

Aside from bacteriophage λ, other prokaryotic DNA templates are assembled into nuclear structures by the *Xenopus* egg extract. Importantly it is possible to assemble on these apparently non-specific DNAs the appropriate nuclear architecture to facilitate the initiation and completion of semiconservative DNA replication. Laskey and colleagues have demonstrated using confocal scanning laser microscopy that replication takes place at discrete chromosomal sites in *Xenopus* sperm nuclei incubated in the egg extract. The number and nuclear distribution of these sites is similar to those observed for somatic cells in tissue culture. Each site (100–300 per nucleus) contains 300–1000 tightly clustered replication forks. Laskey demonstrated that comparable replication structures are assembled on bacteriophage λ DNA in the egg extract. These results establish the important fact that this highly regulated functional chromosome structure can be established *de novo* independent of any precise eukaryotic DNA sequence or pre-existing chromosomal or nuclear structure (Cox and Laskey, 1991).

Summary

The assembly of nuclei *in vitro* does not require any specific eukaryotic DNA sequence, but does require DNA. A nucleus is built from the initial assembly of simple structures to the final assembly of more complex ones. The assembly of chromatin precedes the assembly of a nuclear lamina and nuclear membrane. Not until the nuclear membrane is complete will the nucleus swell to its final volume. Once the complete nucleus is assembled, regulated events such as the initiation of DNA replication can take place, even on prokaryotic DNA templates.

CHAPTER FOUR

How Do Nuclear Processes Occur in Chromatin?

At first sight the folding of DNA into a chromosome presents many impediments to any potential metabolic process requiring access to the double helix. Even though DNA is severely compacted, complex events such as replication, transcription, recombination and repair must occur efficiently in a chromatin environment. Evolution has been remarkably successful in shaping chromatin such that it does not prevent *trans*-acting factors from gaining access to specific DNA sequences or hinder polymerases from progressing along the chromatin fiber. We will see that eukaryotic *trans*-acting factors have evolved to operate in a chromatin environment and that histones have evolved to let them function.

4.1 OVERVIEW OF NUCLEAR PROCESSES

Most regulated events involving DNA offer what appears to be a bewildering complexity of specific DNA sequences (*cis*-acting elements) and proteins (*trans*-acting factors) controlling a particular process. Although the individual proteins and DNA sequences regulating events differ for DNA replication, recombination, repair or transcription, certain general principles apply to each. For eukaryotes, the regulation of transcription has by far the best understood molecular mechanisms; however, much of our insight into the control of metabolic events involving DNA was first established in a variety of prokaryotic systems.

4.1.1 The problem of specificity

The conventional approach to dissecting a complex process follows from the biochemical fractionation of crude extracts *in vitro*. The molecular dissection of the chromatin assembly process can be categorized in this way (Section 3.4). Kornberg and colleagues have explored in some considerable detail the molecular mechanisms controlling the highly regulated initiation of chromosomal replication in *Escherichia coli* (Kornberg, 1988). This event normally occurs once per cell generation at a single site selected from the entire *E. coli* genome (4×10^6 bp). This unique chromosomal origin (*oriC*, 245 bp in length) is recognized by a sequence-specific DNA binding protein, dnaA, that associates with four non-contiguous 9 bp repeats. As the dnaA protein functions to specifically determine the site at which replication will initiate, it can be described as a specificity factor. The protein can interact with individual 9 bp repeats, but only when four are placed in the correct positions relative to each other can a large complex of dnaA protein and the *oriC* DNA sequence be formed. It is this large nucleoprotein complex that is recognized by the other proteins required for replication.

The dnaA protein not only interacts with DNA but also associates through direct protein–protein contacts with other dnaA molecules. This results in a cooperative association of 20–30 dnaA molecules with *oriC*. DNA is wrapped around the complex of dnaA protein rather like it is around the core histones (Section 2.2.2). The dnaA protein–DNA complex formed at *oriC* facilitates a specific duplex opening reaction in an adjacent AT-rich DNA sequence. The dnaC protein mediates the association of the dnaB helicase with this single-stranded AT-rich sequence, unwinding a substantial segment of the double helix, at which replication enzymes such as DNA primase and polymerase begin to act. Interestingly, the protein most analogous to a histone in *E. coli*, the HU protein, facilitates the formation of a functional nucleoprotein complex at *oriC*, as do topoisomerases. The role of these proteins is probably to facilitate the correct topological arrangement of DNA for the subsequent replication events.

A similar pattern of events occurs during the initiation of replication in bacteriophage λ. The O protein interacts as a dimer with four repeated sequences (18 bp). Electron microscopy reveals that a specific nucleoprotein structure is assembled at the λ origin, with

DNA wrapped on the outside (the O-some). Once again other proteins recognize the O-some and initiate the replication process (Echols, 1986; Fig. 4.1). Site-specific recombination by bacteriophage λ also employs proteins (Int) that bind at multiple sites (the att P site) to arrange DNA into a specific nucleoprotein complex (the intasome). A specialized HU protein, IHF (integration host factor) facilitates the recombination process both by assisting Int binding and bybending DNA (Yang and Nash, 1989).

Several general rules follow from these analyses. All of these processes require exceptional precision, as do similar events in eukaryotic cells. Although DNA binding proteins can interact with these specific sites with high affinity ($K_D = 10^{-9}$–10^{-13} M) they also bind DNA non-specifically ($K_D = 10^{-6}$–10^{-3} M). Thus the basis of the exceptional precision of replication and recombination events in *E. coli* is unlikely to follow from the binding of a *single* protein to a *single* DNA site. Instead, as illustrated by the examples, the cooperative interaction of a particular protein (the dnaA or O proteins) with multiple sites over a 200–300 bp region of DNA is required for initiation of these processes. The precise organization of DNA into these complexes is necessary for other proteins to recognize the origin or integration site. This precise organization also requires the mediation of proteins that alter DNA conformation (HU or IHF) and remove topological constraints to the folding of DNA (topoisomerases).

As discussed by Echols (1986), not all metabolic events involving DNA require the level of precision or regulation inherent to chromosomal replication in *E. coli*. Transcription events in *E. coli* do not need a comparable level of accuracy since the occasional erroneous initiation event is unlikely to kill the cell. However, the human genome of 3×10^9 bp is roughly a thousand times greater in size and consequently more complex than that of *E. coli*. It is probable that the human genome contains over a thousand times more low affinity or non-specific protein binding sites than are seen in *E. coli*. This number of sites will, in principle, have to be scanned through non-specific protein–DNA interactions before the correct binding region is found and specific binding occurs. The promoters of eukaryotic genes are not three orders of magnitude less accurately regulated than their prokaryotic counterparts. Therefore, it is safe to assume that the transcription process in eukaryotes has a level of precision comparable to prokaryotic replication or recombination.

The precision of regulated events in eukaryotic cells is determined both by following the prokaryotic paradigm and utilizing multiple high-affinity sequence-specific DNA-binding proteins that recognize multiple related sequence elements, and by masking many of the

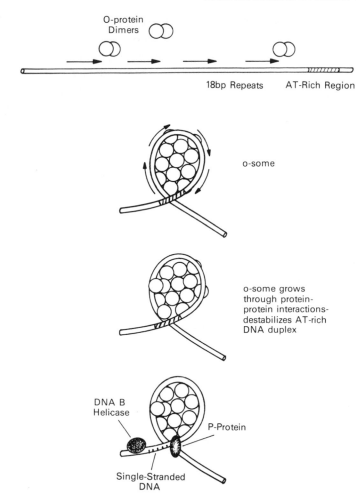

Figure 4.1. Formation of the O-some at the origin of replication of bacteriophage λ.

non-specific sites by folding them into chromatin (Lin and Riggs, 1975, Section 4.2.1). The first examples of multiple binding sites within DNA for particular sequence recognition proteins regulating replication or transcription in eukaryotic cells were determined in viral systems. The SV40 genome has three binding sites for T-antigen at the origin of replication (ORI), a nucleosome-free region in the minichromosome (Section 3.2.2; see Fig. 3.3). Scanning transmission electron microscopy reveals that trimers and tetramers of T-antigen bind at each of these sites (Mastrangelo *et al.*, 1985). Like dnaA, T-antigen also causes significant changes in local DNA structure at the origin (Challberg and Kelly, 1989). Other viral genomes have a similar requirement for multiple binding sites to be occupied by a particular virally encoded protein in order to initiate replication. The nuclear antigen (EBNA-1) protein of Epstein-Barr virus has six specific binding sites in the viral replication origin region. Thus eukaryotic viruses are likely to regulate replication in a comparable way to *E. coli*. In contrast to this tight regulation of replication by sequence-specific proteins, the organization of chromosomal origins of replication in eukaryotic cells is presently unknown, but appears to rely on aspects of chromosomal structure and the nuclear envelope (Section 3.4.4).

The isolation and purification of eukaryotic transcription factors has allowed the definition of multiple sites of interaction for these proteins both in viral DNA and in normal cellular promoters. Perhaps the best-studied example for this class of protein is the role of the transcription factor SP1 in transcriptional regulation. SP1 was originally defined as a promoter-specific transcription factor required for the efficient recognition by RNA polymerase II of the early and late promoters of SV40. SP1 was found to bind at six sites within the SV40 origin region stimulating transcription from both promoters, directed away from the origin (Dynan and Tjian, 1985; see Fig. 3.3). Multiple SP1 binding sites were subsequently defined in many normal cellular genes, including four sites in the mouse dihydrofolate reductase promoter. The significance of these multiple sites is not only the increase in the precision of transcriptional regulation, but also that a potential synergism exists in nucleoprotein complex formation mediated by protein–protein interactions between SP1 molecules. SP1 molecules bound to weak and strong sites help each other to bind to DNA, like the dnaA protein does on interaction with *oriC* (Courey *et al.*, 1989). This synergistic effect extends to SP1 sites some distance (> 1 kb) away from each other (Section 4.1.2). Therefore, the regulation of the specific initiation of transcription at the early and late promoter of SV40 resembles the control of replication initiation in *E. coli*.

Summary
The regulation of replication in *E. coli* requires the specific association of a DNA binding protein with multiple copies of a particular DNA sequence. A large complex of DNA and protein is formed due to cooperative interactions between DNA-bound and -unbound protein molecules. Formation of this complex is facilitated by other auxiliary proteins that alter DNA conformation or remove topological constraints. The regulation of replication and transcription initiation in

eukaryotic viruses and of transcription initiation in chromosomal promoters show many parallels with the control of *E. coli* replication.

4.1.2 Action at a distance

A common feature of eukaryotic transcriptional regulation is its control by two types of DNA sequences containing clusters of *trans*-acting factor binding sites. One type of sequence is generally located just 5′ to the start site of transcription (proximal promoter elements). The other DNA sequence is generally at some distance (several kilobases) away from the start site of transcription (enhancers). Enhancers act in a position- and orientation-independent manner to stimulate transcription. DNA sequences that act in a comparable way to inhibit transcription are known as silencers (Brand *et al.*, 1985). How enhancers work is not known, although they clearly facilitate the binding of *trans*-acting factors to the proximal promoter elements (Mattaj *et al.*, 1985; Weintraub, 1988). It is possible that enhancers act by causing a local unravelling of the chromatin fiber thereby facilitating transcription factor access to DNA (Section 4.2.4). Another popular mechanism is that direct contact occurs between proximal promoter elements and the enhancer through looping-out the intervening DNA (Ptashne, 1986).

Support for the looping-out hypothesis comes from examination of the regulation of transcription in *E. coli* (Gralla, 1991). Most *E. coli* promoters are regulated by the binding of one or two molecules of activator or repressor proteins immediately adjacent to the start site of transcription. However, certain promoters, such as the L-arabinose BAD operon, are regulated by protein–DNA interactions over an extensive region requiring the formation of a DNA loop. For example, a loop of DNA involving over 210 bp of DNA is mediated by the ara C protein leading to the repression of transcription from the adjacent operon (Lobell and Schlief, 1989). The resulting nucleoprotein complex has many similarities to those that regulate replication and recombination (Section 4.1.1). An important limitation to the hypothesis that similar events occur in eukaryotes is that all of the *E. coli* loops are quite small (< 500 bp). In eukaryotes, enhancers can act over kilobases of DNA. It is possible that the removal of eukaryotic enhancers to greater distances is because the intervening DNA can be folded into nucleosomes and the chromatin fiber.

The capacity of the eukaryotic protein SP1 to activate transcription from distant sites by looping-out intervening DNA has been tested. The herpes simplex virus (HSV) thymidine kinase (tk) promoter contains only two SP1 binding sites; however, this is sufficient to provide a measurable stimulation of transcription through the association of SP1. Insertion of six binding sites ~ 1.8 kb upstream of the tk promoter greatly stimulates transcription (a 90-fold induction occurs, Courey *et al.*, 1989). Direct interaction between these two sites was shown to take place by electron microscopy (Mastrangelo *et al.*, 1991; Su *et al.*, 1991). Such a model of nucleoprotein complex formation involving both enhancer and promoter DNA mediated by acommon protein would be consistent with the capacity of enhancers to act even when not physically linked to a promoter. Other experiments have shown that enhancers can stimulate transcription from a promoter on a separate DNA molecule, albeit that the two distinct DNA molecules containing the enhancer or promoter are intertwined and therefore constrained in space (Muller-Storm *et al.*, 1989).

Summary
Many eukaryotic genes are controlled not only at promoters adjacent to the start site of transcription but also by enhancers that can be several kilobases away. Looping-out of intervening DNA to form a common nucleoprotein complex involving both enhancer and promoter DNA sequences might explain the action of enhancers at promoters over great distances.

4.1.3 The basal transcriptional machinery

SP1 is only one example of a plethora of sequence-specific DNA binding proteins that regulate transcription in eukaryotes. These different proteins may be grouped on the basis of structure (Section 4.1.6); however, they all have one thing in common. All of these factors influence the function of the basal transcriptional machinery, in particular the assembly of a nucleoprotein complex that can be recognized by RNA polymerase (Mitchell and Tjian, 1989).

Roeder and colleagues have defined the basic components required for transcription by RNA polymerase II. These include the TFIID protein that binds to the TATA box (normally 30 bp 5′ to the transcription start site in mammals, as much as 100 bp 5′ in yeast). Binding of TFIID to the TATA box facilitates the stepwise association of the other general transcription factors (TF)IIA, B and E (yet more exist that facilitate later steps in the transcription process, TFIIF, S and others) (Pugh and Tjian, 1992). Once the transcription (or pre-initiation) complex is assembled, it can be recognized by RNA polymerase II and the actual process of synthesizing RNA initiated. The binding of TFIID commits a promoter to be transcriptionally active. The TFIID protein remains stably bound to the promoter following transcription initiation and elongation by RNA polymerase II (van Dyke *et al.*, 1988). It follows from this stable association of TFIID with a promoter that events regulating its access to the TATA box are very important. Several experiments with viral and cellular promoter-specific proteins have shown that they facilitate TFIID binding (Sawadogo and Roeder, 1985; Workman *et al.*, 1988; Horikoshi *et al.*, 1988); other proteins can repress transcription by competition with TFIID for binding to the TATA box (Ohkuma *et al.*, 1990).

How promoter-specific initiation factors interact with TFIID and other components of the basal transcriptional apparatus is currently a subject of intense interest and controversy (Carey, 1991). The cloning of TFIID has allowed this issue to be directly addressed (Hahn *et al.*, 1989). It has been proposed, based on the failure of recombinant TFIID to respond to an activator such as SP1, that molecules exist, known as coactivators, that allow the basal transcriptional machinery to respond. Coactivators might establish contacts both with SP1 and with a component of the pre-initiation complex forming a bridge between the two proteins (Pugh and Tjian, 1990; Smale *et al.*, 1990). Certain domains of promoter-specific factors such as anionic regions (Section 4.1.6) have been shown to interact with TFIID (Stringer *et al.*, 1990), although a distinct component of the basal transcriptional machinery TFIIB (which binds subsequent to TFIID) interacts even more avidly with such domains (Lin and Green, 1991). Other studies support an important role for TFIIB indicating that the efficiency with which TFIIB can be sequestered into a pre-initiation complex can be influenced by promoter specific transcription factors (Carey, 1991). Therefore it seems that either limiting TFIID or TFIIB can determine the transcriptional activity of a promoter recognized by RNA polymerase II. Either the binding of TFIID to DNA or a later step in making the transcription complex competent can be regulated.

In contrast to the complexity of transcriptional regulation of the class II genes encoding mRNA, the proteins required to transcribe class III and class I genes are relatively simple. The study of the transcription of 5S RNA and tRNA genes (class III) by RNA polymerase III and of ribosomal RNA genes (class I) by RNA polymerase I has been very informative with respect to the influence of chromatin structure on the initiation of transcription and the consequences of transcription for chromatin structure (Sections 4.2 and 4.3). Much of our insight into transcriptional regulatory process has been established with these genes (Wolffe, 1991b).

Three proteins are required to assemble a transcription complex on a 5S RNA gene that can be recognized by RNA polymerase III. Transcription factor (TF)IIIA is a promoter-specific DNA binding protein that only recognizes the 5S RNA gene specifically, whereas TFIIIC and TFIIIB are proteins required for both 5S RNA and tRNA gene transcription. Unlike class II genes, no enhancer elements are known that can influence the efficiency of transcription complex formation on 5S or tRNA genes. Instead, an interesting hierarchy of transcription factor–DNA and protein–protein interactions between TFIIIA, B and C occur that serve to regulate differential class III gene transcription.

TFIIIA is a simple protein consisting of an array of zinc-finger domains (Section 4.1.6) that interacts with DNA at a site within the 5S RNA gene called the internal control region. Although TFIIIA associates with DNA at a specific sequence, the binding affinity for DNA is low ($K_D = 10^{-9}$ M). However, TFIIIA has to form a complex with the 5S RNA gene before TFIIIC can be sequestered onto the gene. TFIIIC binding can influence the stability with which TFIIIA binds to the 5S RNA gene, with important consequences for gene regulation (Section 4.1.5). Once TFIIIC is bound, the rate-limiting step in transcription complex formation occurs; this is the binding of TFIIIB to the TFIIIA–C 5S DNA complex. RNA polymerase can only recognize the transcription complex, and thus the 5S RNA gene, after TFIIIB has been sequestered. A similar process occurs on a tRNA gene, except that TFIIIA can be dispensed with. Here TFIIIC binds directly and specifically to the promoter elements of the gene. TFIIIB can then associate with the TFIIIC–tRNA gene complex and RNA polymerase can then recognize the gene (Fig. 4.2).

TFIIIA and TFIIIC can be described as assembly factors, since TFIIIB appears to be the only protein directly recognized by RNA polymerase III (Kassavetis *et al.*, 1989). Using *Saccharomyces cerevisiae*, Geiduschek and colleagues were able to show that TFIIIA and C could be removed from a 5S RNA gene (or TFIIIC from a tRNA gene) leaving TFIIIB in place. Under these conditions multiple rounds of transcription initiation by RNA polymerase III could still occur. Surprisingly specific DNA sequences within the 5' flanking regions of class III genes are not essential for the efficient transcription of these genes, yet it is this region that TFIIIB interacts with. Moreover, TFIIIB itself is not a DNA binding protein. This implies that TFIIIA and TFIIIC are not only essential for bringing TFIIIB to the class III gene, but also for activating its DNA binding activity and precisely positioning it at the appropriate place to interact with RNA polymerase III. It is remarkable that TFIIIB binds so tightly to DNA that it can only be

tRNA Gene   5S rRNA Gene

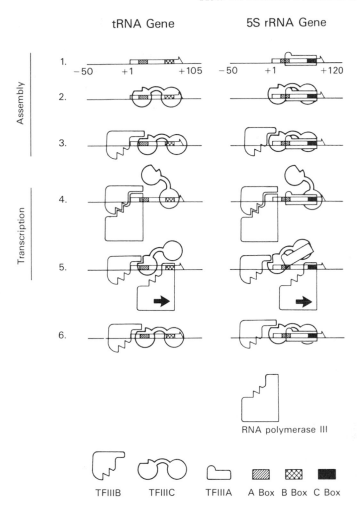

TFIIIB   TFIIIC   TFIIIA   A Box   B Box   C Box

Figure 4.2. Assembly of transcription complexes on tRNA and 5S RNA genes, and a hypothetical mechanism for transcription through the complex without displacement of transcription factors.

Transcription factors, their binding sites and RNA polymerase III are shown in the cartoon.

dissociated by chaotropic agents, yet this binding is activated by protein–protein interactions and is non-specific. TFIIIA and TFIIIC appear to fulfill the same function of directing the association of the protein recognized by RNA polymerase for class III genes, that the large number of promoter-specific transcription factors do for class II genes. Gene regulation therefore concerns any process that influences the sequestration at the promoter of the key transcription factor: TFIIIB for class III genes; TFIIB or TFIID for class II genes (Section 4.1.5).

Several analogies between the transcription of class III, class II and class I genes exist (Paule, 1990). Ribosomal RNA gene transcription requires a single factor (TIF-1) in *Acanthamoeba*, which remains in place through multiple transcription initiation events *in vitro* rather like TFIID and TFIIIB. However, in vertebrates two proteins are required, called upstream binding factor (UBF) and a poorly defined protein known as SL1 (Jantzen *et al.*, 1990). SL1 and UBF interact cooperatively with DNA at the promoter element of class I genes (Bell *et al.*, 1988). This type of cooperative interaction is similar to the stabilization of TFIIIA binding by TFIIIC on the 5S RNA gene. UBF also functions in a similar way to class II transcription factors in that aside from the promoter, it also binds to repetitive sequences 5′ to the ribosomal RNA gene promoter that function as enhancer elements. Ribosomal RNA genetranscription is therefore an excellent example of promoter specificity and regulation being assisted by multiple protein–DNA interactions involving a limited number of proteins and repeated recognition sites (Echols, 1986, 1990).

In contrast to the multiple promoter-specific DNA binding proteins the complex multisubunit RNA polymerases have not been shown to have a role in facilitating the assembly of transcription complexes. The fact that they have high affinity for proteins associated with transcription complexes implies that certain transcription factors may associate with RNA polymerase in the absence of DNA. This has been shown to be the case for the general class III transcription factors TFIIIB and TFIIIC (Wingender *et al.*, 1986) and for the basal class II transcription factor TFIIF (Sopta *et al.*, 1989). Some of these proteins may always be associated with RNA polymerase *in vivo*, so the boundary between being a component of the transcription complex or of the polymerase becomes artificial. After transcription initiation and promoter clearance by RNA polymerase some of the non-DNA binding transcription factors may dissociate, preventing reinitiation by the enzyme until they have rebound (Van Dyke *et al.*, 1988; Hai *et al.*, 1988). The DNA binding proteins remain and may be responsible for the phenomenon of template commitment, a function of a nucleoprotein complex unique to eukaryotes. This is the capacity of a nucleoprotein complex, once assembled, to maintain a particular function such as transcription indefinitely (Section 4.1.4).

Summary

Certain general rules emerge concerning the transcription process relevant to all eukaryotic promoters (Wolffe, 1990b). RNA polymerase (I, II or III) does not recognize naked DNA itself but a complex multiprotein structure including the promoter DNA sequence and the start site of transcription (Fig. 4.3). Multiprotein transcription com-

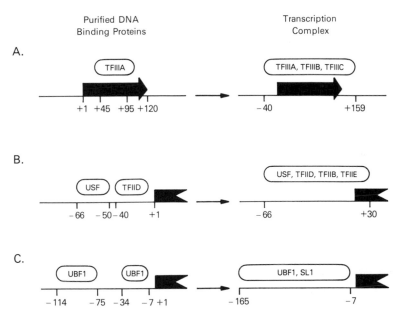

Figure 4.3. Transcription complexes on class I, II and III genes. The regions of promoters associated with transcription factors are indicated for all three classes of eukaryotic gene. The solid bar indicates the gene itself, numbers indicate the boundaries of the DNaseI footprints of either purified DNA binding proteins or of the complete transcription complex. (A) The *Xenopus* somatic 5S RNA gene transcription complex (class III) (Wolffe and Brown, 1988). (B) The adenovirus major late promoter transcription complex (class II) (Van Dyke *et al.*, 1988). (C) The human ribosomal RNA gene transcription complex (class I) (Bell *et al.*, 1988).

plexes undergo a highly ordered assembly process initiated by sequence-specific DNA binding proteins. Non-DNA binding proteins are sequestered by virtue of protein–protein contacts. The association of these additional proteins alters pre-existing DNA–protein interactions both qualitatively (extent or site of contacts) and quantitatively (the affinity of the whole complex for the promoter exceeds that of the DNA binding proteins alone). Over 100 bp of DNA sequence is complexed with multiple proteins on each class (I, II and III) of eukaryotic promoter. Interactions between transcription factors over this length of DNA implies either a precise stereospecific orientation of individual proteins on the surface of the double helix or that considerable flexibility exists in the structure of the bound proteins. One potential function of the extensive protein–DNA and protein–protein interactions at the promoters is to increase the fidelity of transcription initiation. Each specific protein–DNA and protein–protein interaction is a prerequisite for efficient transcription and provides a reference point for RNA polymerase in aligning the initiation site of transcription.

4.1.4 Stable transcription complexes

The stable sequestration of transcription factors onto promoter elements responsible for template commitment (Section 4.1.3) has been described for representatives of all classes (I, II and III) of eukaryotic gene (Brown, 1984). One assay for stable transcription complex formation involves the sequential addition of two genes to an *in vitro* transcription extract. If the first gene added binds all the limiting transcription factors in the extract, then the second gene will not be transcribed. If this is true *in vivo*, then stable transcription complexes might explain the maintenance of distinct patterns of gene activity in a terminally differentiated cell. Experiments involving the injection of genes into a living *Xenopus* oocyte nucleus support this idea. Transcription complexes assembled on a 5S RNA gene (*Xenopus* somatic type) are stable for several days, longer than the lifetime of some cells. Moreover, transcription complexes can be found in chromatin isolated from erythrocyte cells in which there are no longer free transcription factors (Darby *et al.*, 1988; Chipev and Wolffe, 1992). This suggests that transcription complexes, at least on 5S RNA genes, can be stable for several weeks.

The stability of a transcription complex depends on the multiple protein–protein and protein–DNA interactions involved in its assembly (Section 4.2.3). This stability is distinct from the transition in DNA structure required during transcription in prokaryotic systems, when the unstable, closed complex is transformed to the stable, transcriptionally active open complex in which DNA is unwound at the promoter (Hawley and McClure, 1982; Kassavetis *et al.*, 1989). At the eukaryotic transcription complex DNA remains in the double helical form until RNA polymerase initiates transcription. The requirement for many interactions to generate a stable complex affords multiple opportunities for regulating these interactions and thereby modulating gene activity (Section 4.1.5).

An active eukaryotic gene has to maintain transcription factor–DNA interactions through a variety of potentially disruptive events. Transcription, replication and chromatin assembly represent the three major possible mechanisms by which a transcription complex might be disrupted. Chromatin assembly and chromosome compaction can influence transcription complex formation and stability under certain circumstances (Section 4.3.1). If a transcription complex is not stable, changes in chromatin structure may direct the dissociation of transcription factors and the repression of genes (Wolffe, 1989b; Chipev and Wolffe, 1992). However, in general most transcription complexes appear stable to changes in chromatin and chromosomal structure (Section 4.2.2).

An immediate problem in transcribing a class III gene is that all of the essential promoter elements are within the gene sequence (Ciliberto *et al.*, 1983). On a somatic 5S RNA gene, transcription factors associated with these sequences remain stably bound in spite of hundreds of transits by RNA polymerase III (Wolffe *et al.*, 1986). Experiments with bacteriophage SP6 RNA polymerase revealed that the presence of multiple DNA binding proteins coupled together by protein–protein contacts allows individual proteins to anchor the complex to DNA. Transient dissociation of any one contact need not lead to dissociation of the whole complex (see Fig. 4.2). A similar array of contacts might account for the stability of nucleosome structures during transcription (Section 4.3.4).

A very different result is seen when only a single transcription factor is associated with a promoter, as in the case of the *Acanthamoeba* class I ribosomal RNA gene. RNA polymerase passage through this promoter leads to the displacement of the single transcription factor (Bateman and Paule, 1988). This result graphically demonstrates the disadvantage of only a limited number of protein–DNA contacts mediating transcription. Similar results have been obtained when TFIIIA alone is bound to a 5S RNA gene (Campbell and Setzer, 1991). This may also explain the significance of transcription termination sites being placed upstream of the promoters of genes arranged in tandem arrays (e.g. ribosomal RNA genes, McStay and Reeder, 1986). Inhibition of transcription from promoters in the path of a transcribing RNA polymerase is a well-known phenomenon for prokaryotic genes (Adhya and Gottesman, 1982; Horowitz and Platt, 1982).

The capacity to maintain protein–DNA interactions in place during transcription may be an important element of transcription complex structure contributing to the regulation of many genes. Some class II genes are known to have regulatory elements and protein–DNA complexes within either exons or introns (Banerji *et al.*, 1983; La Flamme *et al.*, 1987; Theulaz *et al.*, 1988). The capacity to have stable complexes assembled downstream of the promoter contributes another dimension of flexibility to eukaryotic gene regulation (Schaffner *et al.*, 1988). For example, the adenovirus major late promoter directs RNA polymerase to transcribe a gene which contains five other active promoters. Stable complexes are assembled on both the class III VA genes and class II promoters, these appear to remain in place in spite of the RNA polymerase initiated at the major late promoter moving through them (Berk, 1986). The maintenance of a transcription complex in spite of transcription through it (Wolffe *et al.*, 1986), means that we should not, perhaps, be too surprised to find overlapping transcription units in the eukaryotic genome.

A related problem of maintaining specific protein–DNA interactions associated with a transcription complex occurs when a replication fork passes along a gene. What happens to the transcription factors comprising the complex may have important implications for the inheritance of patterns of gene activity in eukaryotic cells (Brown, 1984). The stable association of *trans*-acting factors with DNA through the replication event would be a simple way of imprinting a particular expression pattern or a promoter through development. Experiments that attempt to test the maintenance of transcription complexes through replication *in vivo* have generally made use of 'enhancer-dependent' promoters. Enhancers facilitate the assembly of transcription complexes at promoters, and in some cases stimulate transcription initiation by over 100 fold (Section 4.1.2). In one experiment, Calame and colleagues established competition for SV40 enhancer factors, after enhancer-dependent transcription had been initiated on a gene (Wang and Calame, 1986). Transcription from the promoter of the gene continued in spite of the competition, indicating that transcription factors were stably sequestered at the promoter. Moreover, replication of the transcriptionally active gene did not inhibit transcription even in the presence of the competitor DNA. This shows that either enhancer action was not required to re-establish the transcription complex once it had been formed, or that the transcription complex on the promoter was stable to replication fork passage.

The problem of template commitment *in vivo* has often been investigated and discussed using immunoglobulin genes as examples. These genes alter their utilization of regulatory elements during lymphoid cell differentiation. The immunoglobulin heavy chain (IgH) enhancer is required to activate transcription from IgH promoters early in B-cell differentiation (Banerji *et al.*, 1983). However, several differentiated B-lymphoid cell lines exist that have deleted the IgH enhancer, but retain normal levels of IgH transcription (Wabl and

Burrows, 1984; Klein *et al.*, 1985). There are several possible explanations for this result. One of these is that the IgH enhancer is required only for the establishment of the IgH promoter transcription complex early in B-cell differentiation, and later on the enhancer can be deleted and the transcription complex will remain in place in spite of cell division. Alternatively, the enhancer is required for maintenance of the IgH gene transcription complex during cell division, but when deleted can be replaced by other regulatory elements (Grosschedl and Marx, 1988). A third explanation might be that some other modification of active chromatin such as demethylation, which can be propagated at the replication fork, might explain continued gene activity (Kelley *et al.*, 1988). Distinguishing between these possibilities is difficult and has not yet been achieved. The maintenance of the transcription complex through replication remains an attractive possibility. However, definitive proof of the stability of a transcription complex through cell division requires the physical structure of transcription complexes assembled on a particular promoter to be analyzed before and after DNA replication.

Evidence that argues against the general maintenance of transcription complexes on all genes during replication comes from *in vitro* experiments in which the physical structure of a 5S RNA gene transcription complex was analyzed before and after replication (Wolffe and Brown, 1986). In contrast to the stability of the transcription complex to transcription, replication fork progression disrupted the complex and displaced transcription factors. No selective advantage existed for rebinding factors to the daughter 5S RNA genes, that had initially had a transcription complex, compared to naked 5S DNA. Constitutively expressed genes such as the 5S RNA gene may not require stability to replication, this property may be restricted to the complexes of tissue-specific genes.

A role for extensive cooperative, protein–protein and protein–DNA interactions in maintaining transcription complex structure following replication therefore remains to be proved. Similar arguments have been made for the maintenance of specific chromatin structures, once again without proof (Section 4.3.2). It would be particularly attractive if some aspect of the nucleoprotein structure of a regulatory element might be maintained, thereby providing a molecular explanation for the establishment and maintenance of stable states of gene expression during embryonic development (Brown, 1984; Weintraub, 1985). As we have discussed, examples such as X-chromosome inactivation suggest that chromatin structure can imprint a state of gene expression on a chromosome that can be maintained through DNA replication and cell division (Section 2.5.5).

The alternative to this type of imprinting is a continual regulation of a state of differentiation that is quite plastic and easily influenced by changes in the abundance of individual *trans*-acting factors that activate or repress genes (Sections 3.1.1 and 3.1.2).

Summary
Transcription complex structure may contribute to several important features characteristic of eukaryotic gene expression. The stable sequestration of transcription factors onto a gene by virtue of cooperative interactions between individual factors can explain the terminal differentiation of a cell type and the stability of a pattern of gene activity over long periods of time. This may be helped by the stability of a complex to processive enzyme complexes such as DNA and RNA polymerases and to the compaction of a gene in chromatin. In particular, the commitment of a gene to a continued state of activity in a given cell lineage might be explained by cooperative interactions between components of a transcription complex and maintenance of the structure through DNA replication and cell division. However, the limited existing experimental evidence supports the disruption of such complexes and the reassembly of chromatin structure *de novo* after every replication event.

4.1.5 Regulation of gene activity

Stable transcription complexes may allow a gene to be active indefinitely, however, many gene systems are regulated. For example, a gene that needs to be inactivated during development might make use of a transcription complex that is unstable. The gene would be active when transcription factors were present at high concentrations but inactive when levels fell below a certain threshold. This type of regulation is seen with the 5S RNA genes of *Xenopus laevis*. During embryogenesis the oocyte 5S RNA genes are turned off, whereas the somatic 5S RNA genes remain active (Wormington and Brown, 1983; Wakefield and Gurdon, 1983). Unlike the situation in *Saccharomyces*, the *Xenopus* transcription complexes depend on the interactions of TFIIIA with TFIIIC with the 5S RNA gene for their stability. Moreover, transcription complexes appear to retain TFIIIA and C *in vitro* and *in vivo* (Wolffe and Morse, 1990; Chipev and Wolffe, 1992). Interestingly, TFIIIA and TFIIIC bind rapidly to the 5S RNA gene, and are therefore the proteins that might have to compete with histones for access to DNA regulatory elements (Section 4.3.1). A reduction in transcription factor (TFIIIA and C) concentration during embryogenesis leads to the selective dissociation of oocyte 5S RNA gene transcription complexes. Furthermore, chromatin assembly both prevents transcription factors reassociating with the oocyte 5S DNA and may direct the dissociation of transcription factors from genes leading to repression (Schlissel and Brown, 1984; Wolffe, 1989b; Section 4.2.3).

The DNA sequence differences responsible for this differential regulation of the 5S RNA genes appear to consist of only three base pairs. However, two transcription factors, TFIIIA and C, bind to this region of the gene (Pieler *et al.*, 1987; Wolffe, 1988). Changes in the binding affinity of each of the two proteins amplify their individual effect on complex stability. A difference in gene activity of over 1000 fold can therefore be explained simply by differences in the stability of protein–protein and protein–DNA interactions in oocyte or somatic 5S RNA gene transcription complexes (Fig. 4.4).

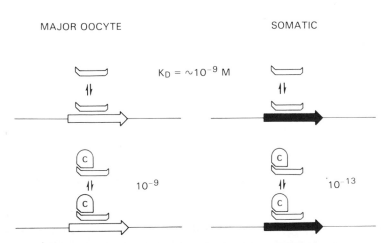

Figure 4.4. Combinatorial interaction of TFIIIC and TFIIIA with oocyte and somatic 5S RNA genes discriminates between them. TFIIIA (open box) binds with equivalent affinity ($K_D = \sim 10^{-9}$ M) to both oocyte (major variant) and somatic 5S RNA genes. Oocyte genes are shown as open horizontal arrows whereas somatic genes are shown as solid horizontal arrows. In the presence of TFIIIC (C), TFIIIA binding to the oocyte 5S RNA gene is unchanged, whereas a very stable complex ($K_D = 10^{-13}$ M) is formed with the somatic 55 RBA gene.

Many eukaryotic genes require enhancers for maximal gene activity and are regulated through changes in protein–DNA interactions at these sequences (Serfling *et al.*, 1985). Although the precise mechanism of enhancer action is not understood, there are many similarities in the assembly of the nucleoprotein structures at both eukaryotic promoters and enhancers. Both complexes are made up of multiple sequence elements, each of which binds a cognate transcription factor (Zenke *et al.*, 1986; Wildeman *et al.*, 1986; Fromental *et al.*, 1988). Both may require stability to either transcription or replication (Schaffner *et al.*, 1988; Wang and Calame, 1985). Stable enhancer complexes may be important in maintaining tissue specificity, even though the activity of particular genes may change (Choi and Engel, 1988). Unstable enhancer complexes are important in regulating gene activity. For example, the mouse mammary tumor virus enhancer is only active when glucocorticoid receptor is bound (Yamamoto, 1985). Removal of the steroid hormone results in transcriptional inactivation, indicating that the glucocorticoid receptor is required for both establishment and maintenance of enhancer mediated effects. The glucocorticoid receptor is known to exert its enhancer effect through specific chromatin structures (Section 4.2.3).

Many DNA binding proteins are shared between enhancer sequences and promoters (Falkner and Zachau, 1984; Bienz and Pelham, 1986; Evans *et al.*, 1988), conceivably non-DNA binding proteins will be also be shared. If the DNA binding protein itself does not bind cooperatively like SP1, the interaction of a non-DNA binding protein with a DNA binding protein at two sites may provide a simple explanation for the possible looping between enhancer and promoter elements (Section 4.1.2). The distinction between transcription complexes and enhancer complexes may, in fact, be artificial. A single structure combining both elements affords much greater possibilities for each of the potential functions discussed above. For instance, one reason for the separation of enhancers and promoters on DNA over extensive distances may be that any one structure might be disrupted by DNA replication, while the other would remain intact (Fig. 4.5). If protein binding to one sequence element influences the binding of proteins to the other, then the intact nucleoprotein complex might facilitate the reformation of the disrupted one (Wolffe, 1990b).

Summary

Eukaryotic transcription complexes have the essential role of directing the accurate and efficient initiation of transcription by RNA polymerase on a particular gene. The focus of much current research in

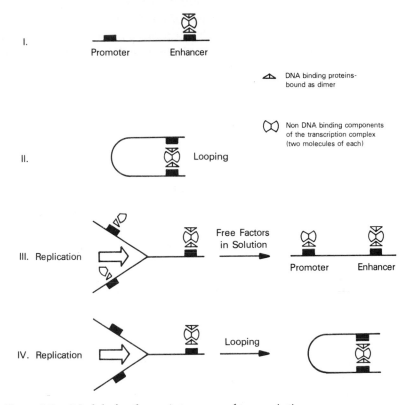

Figure 4.5. Models for the maintenance of transcription complexes through replication.

The regulatory regions of a gene are shown as a promoter and an enhancer (bars). Open boxes are DNA binding or non-DNA binding components of a transcription complex. (I) In this case similar factors are shared between enhancers and promoters. (II) The sequestration of transcription factors onto the promoter (and its activation) is facilitated by looping of the intervening DNA between enhancer and promoter. Common DNA binding proteins associate with both the enhancer and promoter. The bound proteins interact with a common non-DNA binding protein. (III) DNA replication disrupts the transcription complex on the promoter, splitting it in half. The remaining transcription factors can sequester free factors from solution generating a complete transcription complex on each daughter chromatid. In this case the enhancer is required for establishment, but not maintenance of the transcription complex. (IV) Alternatively DNA replication again disrupts the complex on the promoter, displacing transcription factors; however, because of the distance between enhancer and promoter, the enhancer complex remains intact. DNA looping can establish a new transcription complex on the promoter. Here, the enhancer is required for both establishment and maintenance of the transcription complex through cell division.

molecular biology lies in understanding the regulation of the frequency with which RNA polymerase initiates transcription. The multiplicity of proteins and protein–DNA interactions involved in assembling transcription complexes of different stabilities afford many opportunities for regulating complex structure and therefore transcription initiation itself. For example, this could occur through combinatorial effects of multiple proteins binding to particular DNA sequences or by transiently associating and dissociating a particular transcription factor.

4.1.6 Sequence-specific DNA binding proteins

It has been recognized that proteins interacting specifically, or non-specifically with nucleic acids fall into several distinct structural classes dependent on what motifs are present in their amino acid sequences (Johnson and McKnight, 1989). In addition, the motifs that can be discerned in the peptide sequence often represent distinct domains or modules of structure (Frankel and Kim, 1991). As such they can be interchanged using molecular genetics (or evolution) to create new proteins with new functions.

In several instances the motifs present in eukaryotic transcription factors have been shown to adopt highly ordered conformations. In general, these either represent DNA binding or multimerization domains. Well-characterized DNA binding domains include the helix-turn-helix found in homeodomain proteins. This is a unit of three α-helices that binds DNA primarily through contacts in the major groove (Otting *et al.*, 1990; Kissinger *et al.*, 1990; Fig. 4.6). The first DNA binding domain recognized in eukaryotic proteins was the zinc-finger. This structure consists of a two stranded β-sheet and an α-helix in which a single zinc ion is tetrahedrally coordinated to two cysteine and two histidine side-chains (Lee *et al.*, 1989; Klevit *et al.*, 1990; Pavletich and Pabo, 1991). Like the helix-turn-helix proteins, the zinc finger α-helix recognizes specific base pairs in the major groove (Fig. 4.6). A modification of the basic zinc-finger domain is found in the glucocorticoid and estrogen receptors. This domain contains two α-helices that tetrahedrally coordinate two zinc ions through cysteine side-chains. One α-helix from each monomer of the receptor dimer is believed to interact with the major groove of DNA (Hard *et al.*, 1990; Schwabe *et al.*, 1990). The different DNA binding motifs also have distinct requirements for sequence-specific recognition of B-form DNA in solution. These requirements may change when DNA is distorted through interaction with histones.

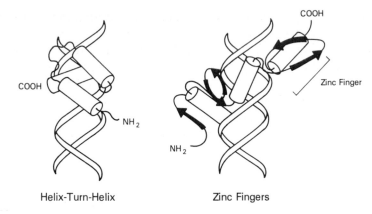

COOH

COOH

Zinc Finger

NH₂

NH₂

Helix-Turn-Helix   Zinc Fingers

Figure 4.6. Binding of helix-turn-helix and a three zinc-finger protein to DNA via contacts in the major groove. α-Helical regions of the protein are shown as cylinders and β-sheet regions as opposed arrows.

In contrast to our rather detailed knowledge of protein–DNA inter-actions, relatively little is known about protein–protein contacts. The only dimerization domain characterized structurally is the leucine zipper. This is a two-stranded, parallel coiled-coil structure (Rasmus-sen *et al.*, 1991). The multiplicity of domains found in different pro-teins presumably reflects the opportunity for multiple independent protein–DNA and protein–protein contacts, each of which might be altered with important consequences for gene regulation (Section 4.1.5). It is also possible that nature has found multiple solutions to a single problem.

Several other multimerization and DNA binding domains have been proposed but their structures remain unknown, these include the helix-loop-helix motif and the cold-shock domain of the Y-box proteins (Murre *et al.*, 1989; Tafuri and Wolffe, 1990). The Y-box proteins contain the most evolutionarily conserved nucleic-acid bind-ing structure yet defined, a block of 70 amino acids that has an identity of 50% between *E. coli* and man. Transcription factors also contain regions of primary amino acid sequence that are involved in their stimulatory action on the transcription process. These include regions rich in acidic amino acids, proline or glutamine (Mitchell and Tjian, 1989). The structure of these regions is unknown. They have been proposed to interact with components of the basal transcrip-tional machinery either directly or through coactivators (Section 4.1.3). Although the activation domains are believed to function through other non-histone proteins they might also function through the modification of chromatin or nucleosome conformation (Section 4.2.2).

Summary
Trans-acting factors generally have a modular structure with distinct DNA binding and protein–protein interaction domains. DNA binding is generally through sequence-specific contacts in the major groove. Their modular structure facilitates multiple independent interactions with DNA and protein, each of which might be regulated to control a process.

4.1.7 Problems for nuclear processes in chromatin

A consideration of the complexity of nuclear processes suggests that the formation of the large nucleoprotein complexes required to control these events might be incompatible with the folding of DNA into chromatin and the chromosome. How can several regions of 100–500 bp exist, each requiring the association of multiple DNA binding proteins to facilitate DNA replication, recombination, repair and transcription; when this DNA may also be wrapped around the core histones and folded into the chromatin fiber? Moreover, how can DNA polymerase and RNA polymerase progress through arrays of nucleosomes and the chromatin fiber even if access to the DNA duplex is achieved? Methodologies to approach these questions have only recently become available. Potential molecular mechanisms to explain the access of *trans*-acting factors to DNA and the progression of polymerases through the chromatin fiber are beginning to be uncovered (Sections 4.2.3 and 4.3).

Since the packaging of DNA into nucleosomes and the chromatin fiber had been thought to remove DNA from any process of interest in the nucleus, experimental analysis has focused either on the regulation of naked DNA templates *in vitro* or on templates uncharacterized with respect to chromatin structure through transient transfection of cells (Section 3.2.2). Results obtained through these analyses are increasingly seen to be oversimplifications of the subtlety and complexity with which genes are regulated in their natural environment – the chromosome (Section 4.2.3). Progress in several experimental systems has clearly shown that promoter elements are specifically organized within and between nucleosomes, and the regulation of a gene depends upon the organization of DNA in a chromatin template (Simpson, 1991). It has also been conclusively demonstrated that nucleosomes, including histone H1, are present on the majority of transcribed genes (Morse, 1992). Replication obviously has to duplicate both DNA and nucleoprotein complexes in order to create two chromosomes out of one. Understanding how these transcription and replication events occur in the context of chromatin will be seen to have regulatory significance important for all nuclear processes.

Summary

Trans-acting factors form large complexes with DNA in spite of the many apparent obstacles due to the concomitant assembly of chromatin. Likewise, processive enzyme complexes function effectively in a chromatin environment. How these events occur in chromatin structures is only beginning to be understood. Understanding the molecular processes that overcome the many apparent impediments to function should uncover regulatory mechanisms unique to eukaryotes.

4.2 INTERACTION OF *TRANS*-ACTING FACTORS WITH CHROMATIN

The difficulties inherent in having non-histone proteins gain access to a histone-covered template were recognized even before the nucleosome model was developed. Experimental approaches to this problem have continually been refined as first non-histone proteins were purified and their binding sites on DNA defined. More recently methodologies for determining specific chromatin structures have been developed. There has been a gradual trend from studying non-specific DNA–protein interactions towards recognition of the role of specific chromatin structures in mediating the function of *trans*-acting factors.

4.2.1 Non-specific interactions

Our knowledge of the accessibility of non-histone proteins to DNA in chromatin has progressed slowly. It has long been known that RNA synthesis using bacterial or bacteriophage RNA polymerase is more efficient from naked DNA than from chromatin, and that the histones are responsible for this inhibition (Georgiev, 1969). These observations led to the idea that DNA was uniformly coated with histones that prevented RNA polymerase from reaching the template. The first experiments to suggest that DNA was not uniformly covered with histones were those of Felsenfeld and colleagues (Clark and Felsenfeld, 1971). Polylysine precipitation of DNA, naked or as chromatin, revealed that as much as 50% of the DNA in chromatin was accessible to the polycation and therefore, presumably naked. This number was very similar to the amount of DNA that could be made acid soluble by nucleases. Clearly some DNA sequences in chromatin were more accessible than others.

The development of the nucleosome concept (Section 2.2) led investigators to explore the relative accessibility to DNA binding proteins of linker DNA compared to that DNA tightly associated with the core histones (core DNA). From the initial definition of the nucleosome through the action of endogenous nucleases, linker DNA is by definition more readily cleaved by these enzymes. Bacterial and bacteriophage RNA polymerases do not transcribe eukaryotic genes with any specificity; however, they will initiate transcription at AT-rich sequences resembling natural prokaryotic promoter elements (Maryanka *et al.*, 1979; Pays *et al.*, 1979). These enzymes have been very useful in assessing the relative accessibility of core versus linker DNA in chromatin. Early studies suggested that DNA in the nucleosome is not accessible to *E. coli* RNA polymerase (Cedar and Felsenfeld, 1973; Williamson and Felsenfeld, 1978). A detailed analysis by Gould and colleagues revealed that linker DNA was more accessible than core DNA to *E. coli* RNA polymerase (Hannon *et al.*, 1984). Titration of linker DNA availability through the addition of histone H1 revealed a rapid decline in accessibility, probably reflecting not only occlusion of linker DNA but also folding of the chromatin fiber. This extensive occlusion of DNA through relatively small changes in linker histone concentration may have significant consequences for the access of other *trans*-acting factors (Section 4.2.3).

These studies were extended to the problem of how a DNA binding protein (*E. coli* RNA polymerase) might search for its binding sites in a nucleosomal array (Hannon *et al.*, 1986). Surprisingly this search occurred with equivalent efficiency in both naked DNA and chromatin that had been depleted of histone H1. Two mechanisms have been envisaged for such a search, either 'sliding' of the DNA binding protein from site to site or 'hopping' between sites (Berg *et al.*, 1981). As *E. coli* RNA polymerase was known not to be able to slide efficiently or progress through nucleosomes (Section 4.3), it was concluded that the enzyme was able to hop between sites efficiently in a chromatin template. These sites are the regions of relatively accessible linker DNA. Removal of histones H2A/H2B from chromatin increases the accessibility of DNA to RNA polymerase even more (Baer and Rhodes, 1983; Gonzalez and Palacian, 1989). In the chromosomal context the search by RNA polymerase for binding sites is probably an accurate reflection of the search of *trans*-acting factors for recognition sequences. As we will discuss, eukaryotic RNA polymerases recognize transcription complexes not naked DNA (Sections 4.1.3 and 4.1.4).

Several interesting biological examples exist of changes in RNA polymerase accessibility to chromatin through development. Brown and colleagues were able to document that the normal somatic form of histone H1 was responsible for maintaining the repression of certain types of class III genes in *Xenopus* somatic cells (Schlissel and Brown, 1984). This repressed state is established gradually during development as the amount of somatic histone H1 increases in chromatin (Wolffe, 1989b, Section 2.5.1). It is interesting that the accumulation of histone H1 correlates with a general decline in the accessibility of RNA polymerase III to DNA (Andrews *et al.*, 1991; Wolffe, 1991a; Fig. 4.7). It is likely that this decline in access to *trans*-acting factors is due to changes in chromatin structure as rapid cell division events cease and a normal cell cycle is imposed.

Experiments with 'non specific' prokaryotic RNA polymerases were responsible for the first demonstration that chromatin structure over a gene isolated from tissues in which the gene was active, differed from that in which it was repressed. These results followed from the relatively easy access of these polymerases to the DNA of transcriptionally active chromatin. Similar results were later obtained using nucleases (Section 4.2.4). Unfortunately, the specificity of transcription was not improved upon using purified eukaryotic RNA polymerases (I, II and III). No eukaryotic RNA polymerase faithfully transcribes specific genes using purified DNA templates. Roeder and colleagues were responsible for the major demonstration that either a natural chromatin template isolated from a cell nucleus or a template reconstituted with transcription factors was necessary for recognition of a genes by RNA polymerase (Parker and Roeder, 1977). At this point the focus of research on gene regulation shifted from the properties of the chromatin template to the properties of the promoter-specific transcription factors (Section 4.1.3).

Summary

Prokaryotic polymerases have been very useful in defining the accessibility of DNA in chromatin to other DNA binding proteins such as *trans*-acting factors. Like nucleases they preferentially recognize

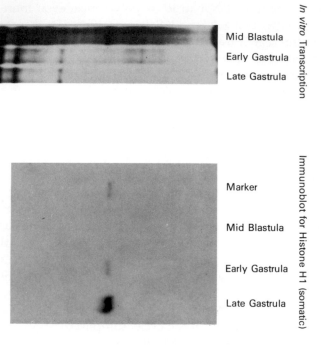

Figure 4.7. The decrease in transcription by RNA polymerase III correlates with the increase of somatic histone H1 in *Xenopus* embryonic chromatin.

Transcription *in vitro* of embryonic chromatin isolated from different developmental stages. Radioactive transcripts were resolved on a denaturing acrylamide gel (specific transcripts are indicated). An immunoblot of *Xenopus* somatic histone H1 in these different chromatin preparations is also shown (see Wolffe 1989b for details).

accessible linker DNA between nucleosomes rather than DNA wrapped around the histone core. They also associate selectively with chromatin that is transcriptionally active *in vivo*. Unlike nucleases they do not destroy the template. Chromatin prevents access of RNA polymerase to DNA; however, the protein can effectively search out binding sites within exposed linker DNA by 'hopping' between sites.

4.2.2 Specific *trans*-acting factors and non-specific chromatin

The availability of *in vitro* transcription systems employing both specific *cis*-acting elements and *trans*-acting factors for the initiation of transcription by RNA polymerase led to a number of experiments in which the influence of chromatin structure on transcription was investigated. A popular experiment with a long history has been to mix a DNA template with histones or a nucleosome assembly system and then to ask whether transcription could still occur. This experiment is responsible for the general belief in the repressive nature of histone–DNA interactions, since the usual result is that the addition of histones inhibits the given process. Although some investigators have undertaken numerous experimental controls to eliminate artifacts, there are often several possible explanations for the observed inhibitory effects that must be excluded.

DNA can precipitate or aggregate following a non-specific association with histones. For example, linker histones (histone H1 or H5) are notorious for forming aggregates on DNA sometimes causing precipitation (Jerzmanowski and Cole, 1990). Most investigators attempt to exclude this possibility by examining the supercoiling or micrococcal nuclease cleavage patterns of their DNA template after nucleosome assembly. Each nucleosome should introduce one negative superhelical turn in the presence of topoisomerase into a closed circular DNA molecule (Fig. 4.8), and protect approximately 146 bp of DNA from micrococcal nuclease (Section 2.2). If these events occur, some fraction of the template must be in solution and contain nucleosomes. However, subnucleosomal particles and proteolyzed nucleosomes will also supercoil DNA and protect it from nucleases, therefore these assays give no guarantee of nucleosome integrity (Section 2.2.4). Unfortunately, it is all too easy to detect a few superhelical turns or to detect a single nucleosome length fragment of DNA after micrococcal nuclease, but difficult to prove that the DNA molecule is efficiently (> 50%) assembled with nucleosomes.

Another difficulty with these reconstitution experiments is excluding the possibility that the template is also associated with non-specific DNA binding proteins. These proteins are often present in crude nucleosome assembly extracts and might also occlude *cis*-acting sequences. Even if efficient nucleosome assembly does occur, the various systems do not always position nucleosomes as found in the chromosome (Section 4.2.3), nor do they always correctly space nucleosomes as would be found *in vivo* (Section 3.4.1). These discrepancies might be explained by the fact that *in vivo*, chromatin assembly is coupled to the replication of DNA, special chromatin assembly factors are employed, the histones are post-translationally modified, and nucleosome assembly is staged (Section 3.3). Thus it is not surprising that the prior association of unmodified histones with DNA under artificial conditions often leads to repressive effects. The physiological

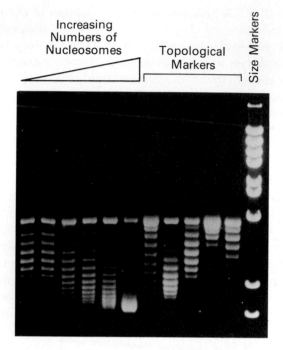

Figure 4.8.   Introduction of supercoils into DNA with increasing numbers of nucleosomes.
An agarose gel resolving topoisomers of a small closed circular plasmid is shown. DNA in the far left lane has very few nucleosomes, more are added (by salt–urea dialysis) from left to right as indicated. The DNA is relaxed with topoisomerase I before deproteinization and resolution on the gel. Markers are also resolved so that the number of nucleosomes can be accurately counted.

significance of the repression may be questionable. With these reservations in mind it is possible to critically evaluate the large body of data on specific transcription factors and non-specific chromatin.

The pioneering studies of Brown, Roeder and colleagues in developing *in vitro* transcription systems led class III genes to be most intensively studied by this type of analysis (Section 4.1.3). Several experiments showed that mixing histones with class III genes would prevent their transcription (Bogenhagen *et al.*, 1982; Gargiulo *et al.*, 1985); however, the organization of the chromatin template was not characterized. In contrast, it was found that *in vivo*, correctly spaced nucleosomes actually correlated with efficient 5S RNA gene transcription (Weisbrod *et al.*, 1982; Gargiulo and Worcel, 1983). An important conclusion was that the transcription factors had to gain access to DNA before the histones if transcription was to occur (Gottesfeld and

Bloomer, 1982; Bogenhagen *et al.*, 1982; Gargiulo *et al.*, 1985). Subsequent studies have shown that a complete transcription complex (TFIIIA, B and C) assembled onto a 5S RNA gene is more resistant to chromatin-mediated repression than the TFIIIA–5S RNA complex alone (Felts *et al.*, 1990; Tremethick *et al.*, 1990). The experiments discussed to this point use chromatin assembly systems that are deficient in histone H1. Reconstitution of histone H1 into chromatin such that the interaction alters nucleosome spacing (and thus is likely to be physiologically relevant) allows inhibition of 5S RNA gene transcription at a lower density of nucleosomes per length of DNA sequence than is otherwise required. In the absence of histone H1, very high nucleosome densities (one every 160–180 bp) are required to repress transcription (Shimamura *et al.*, 1988, 1989; Clark and Wolffe, 1991; Fig. 4.9). Chromatin appears to be repressive, yet if

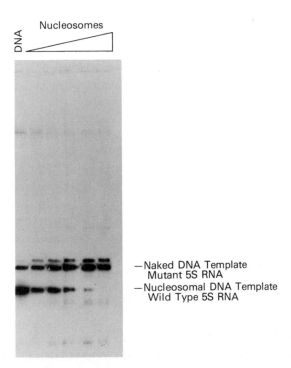

—Naked DNA Template
 Mutant 5S RNA
—Nucleosomal DNA Template
 Wild Type 5S RNA

Figure 4.9. The degree of repression of 5S RNA gene transcription depends on the number of nucleosomes reconstituted on the plasmid.
Radioactive transcripts are shown from a mixture of wild-type 5S DNA reconstituted with increasing numbers of nucleosomes (nucleosomal DNA) and naked mutant 5S DNA which generates a longer transcript (naked DNA) (for details see Clark and Wolffe, 1991).

transcription factors gain access to DNA first, RNA polymerase III has no problems finding the transcription complex in chromatin.

A similar series of experiments have examined the effect of nucleosome assembly on the transcription of class II genes. Luse, Matsui and colleagues showed that prior chromatin assembly restricted transcription from the adenovirus type 2 major late promoter. A major problem of this type of experiment is that only a small percentage (< 5%) of templates are actually transcribed. In contrast class III gene transcription can be very efficient with over 50% of genes in transcription complexes (Clark and Wolffe, 1991). As long as the promoter elements were free it was possible to demonstrate that the small number of active class II genes were in fact assembled into chromatin, suggesting that nucleosomes did not completely inhibit RNA polymerase from transcribing once it had initiated the process (Knezetic and Luse, 1986; Knetzetic *et al.*, 1988; Matsui, 1987). As with class III genes, transcription complexes formed prior to chromatin assembly resisted repression. Roeder and colleagues extended this analysis to suggest that TFIID binding alone was sufficient to relieve the inhibition of transcription due to nucleosome assembly. These experiments have, however, used crude assembly extents supplemented with mixtures of histones including histone H1, and analysis of the resulting chromatin assembly has not been extensive (Workman and Roeder, 1987; Meisterernst *et al.*, 1990). An important point is that promoter-specific factors that stimulate TFIID binding to promoters facilitate transcription of the promoter in the face of whatever is inhibiting transcription in the extract, including chromatin assembly (Workman *et al.*, 1988, 1990). In contrast to this simple race for binding to the promoter, Wu and colleagues have shown that TFIID binding is insufficient for transcriptional activation of the *Drosophila* hsp 70 promoter (Becker *et al.*, 1991). However, a 'potentiated' chromatin template is assembled that can respond to the presence of an 'activated' heat shock transcription factor (HSTF). It has been suggested that HSTF requires TFIID in order to bind to nucleosomal templates (Taylor *et al.*, 1991).

Kingston and colleagues have begun to examine the influence of particular activation domains, especially regions rich in acidic amino acids, in regulating the transcription process in chromatin (Section 4.1.6). It was suggested that the presence of a transcription factor containing these regions could perturb nucleosome structure in the local region (< 100 bp) around the factor binding site (Workman *et al.*, 1991). Unfortunately, none of these studies document the biochemical composition of the assembled chromatin, or whether or not nucleosomes within the vicinity of the promoters are positioned. We will see later that such considerations are very important for gene regulation *in vivo* (Section 4.2.3).

As observed for the repression of class III genes by prior chromatin assembly, supplementation of a nucleosome assembly system with histone H1 more efficiently inhibits class II gene transcription (Laybourn and Kadonaga, 1991). Interestingly, histone H1 is a contaminant of many crude *in vitro* transcription extracts, but not those from *Xenopus* oocytes or eggs which are deficient in histone H1 (Section 3.4.2). Kadonaga and colleagues showed that the action of several promoter-specific DNA binding proteins was to relieve this non-specific inhibitory process. Thus *in vitro* antirepression of transcription is as important as the actual activation process (Croston *et al.*, 1991). The *in vivo* significance of these observations is unclear at this time. Similar results have been obtained for eukaryotic replication origins (Cheng and Kelly, 1989).

Summary
Any results involving non-specific chromatin structures and specific *trans*-acting factors should be treated with caution, especially if the organization of the chromatin template is not documented. General conclusions from a large number of experiments are that prior assembly of a template with nucleosomes inhibits *trans*-acting factor access to DNA, whereas if the factors bind first, subsequent chromatin assembly will not be inhibitory. RNA polymerase can seek out a transcription complex in a chromatin template without any problem. Presumably the complex is a highly visible landmark in an invisible chromatin background.

4.2.3 Specific *trans*-acting factors and specific chromatin

Although the association of *trans*-acting factors with specific DNA sequences was readily accepted, the significance of the 'sequence specific' organization of DNA into nucleosomes has taken longer to be acknowledged (Sections 2.2.5 and 3.2.3). Most investigators accept that there is no logical necessity to organize the vast majority of DNA into chromatin structures that have specific DNA sequences organized in a precise way (Simpson, 1991). However, it is also clear that nucleosomes are positioned around DNA sequences with important functional roles. Formation of such specific chromatin structures is true for the vast majority of genes for which the appropriate assays have been carried out to assess nucleosome positioning. Incorporation of *cis*-acting elements into a positioned nucleosome has important consequences for its accessibility of DNA to *trans*-acting factors. DNA in the nucleosome is highly bent and the helical periodicity of the double helix changes from an average of 10.5 bp/turn to 10.2 bp/turn. Thus, not only is one face of the DNA helix occluded by the histone core, but DNA has an entirely different structure from that in solution (Section 2.2.3).

Martinson and colleagues were the first to examine the issue of *trans*-acting factor access to a specific DNA sequence incorporated into a nucleosome. Their experiments made use of a prokaryotic repressor (the *lac* repressor) and a 144 bp DNA fragment containing binding sites for the repressor. This short DNA fragment is able to assemble a single nucleosome structure following reconstitution with equimolar amounts of the four core histones. The *lac* repressor is a helix-turn-helix protein that binds to B-form DNA on one side of the double helix in the major groove. Based on sedimentation studies the authors concluded that both *lac* repressor and the histone octamer could simultaneously occupy the same DNA fragment. This implied that a triple complex of the DNA, histone proteins and *trans*-acting factor formed (Chao *et al.*, 1980a, b). Unfortunately, these early studies lack the resolution necessary to be absolutely sure of either nucleosome positioning or specific association of the DNA binding proteins; however, they clearly demonstrated the correct approach to this problem.

Transcription factor (TF)IIIA was the first sequence-specific eukaryotic DNA binding protein to be purified. This protein (38 kDa) consists of a chain of zinc-fingers which bind in clusters over the 50 bp internal control region of the 5S RNA gene (Section 4.1.3). Simpson had demonstrated that 5S RNA genes contain strong nucleosome-positioning sequences (Simpson and Stafford, 1983; FitzGerald and Simpson, 1985; Section 2.2.5). In an important experiment, Rhodes examined the interaction of TFIIIA with a *Xenopus* 5S RNA gene that had been incorporated into a nucleosome. It was found that TFIIIA could form a triple complex with the 5S RNA gene and histones. Subsequent experiments have shown that the efficiency of triple complex formation depends on the stoichiometry and post-transcriptional modification of the histone proteins (Hayes *et al.*, 1992). If the nucleosome is deficient in histones H2A/H2B, more DNA is free at the edge of the nucleosome for interaction with TFIIIA. Unlike the lac repressor, which can interact with DNA actually within a nucleosome, TFIIIA binds primarily to free DNA at the edge of the nucleosome, not to DNA actually contacting the histones. Interestingly, acetylation of the core histones facilitates the binding of TFIIIA to the 5S

RNA gene within the nucleosome (Lee *et al.*, 1992). Since acetylation of the core histones does not apparently alter DNA conformation in the nucleosome directly, or the position of the nucleosome relative to the 5S RNA gene, it is likely that a small change in the stability of histone–DNA contacts allows TFIIIA to bind (Section 2.5.2).

The failure of TFIIIA to bind to a 5S RNA gene associated with a complete unmodified octamer of histones to form a positioned nucleosome, is consistent with functional studies *in vitro* (Shimamura *et al.*, 1988; Morse, 1989; Clark and Wolffe, 1991). Using the yeast minichromosome system it has been shown that even yeast nucleosomes will position on a 5S RNA gene *in vivo* (Section 3.2.3). This sequence-directed positioning is not surprising since the 5S RNA gene-positioning sequence is one of the strongest natural ones known, probably because of inherent DNA curvature (Shrader and Crothers, 1989; Hayes *et al.*, 1990; Section 2.2.5). In the natural chromosomal context neither the *Xenopus borealis* or *Xenopus laevis* 5S RNA genes (somatic type) position nucleosomes (Chipev and Wolffe, 1992). However, it is possible that as transcription factors compete for association with a 5S RNA gene, the positioning of a nucleosome or subnucleosomal particle is only transiently important. The positioning of a nucleosome away from the key promoter elements might occur immediately following DNA replication, when competition with transcription factors for association with the gene occurs (Section 4.3). Interestingly, repression of the oocyte 5S RNA genes that bind transcription factors weakly (Section 4.1.5) appears to depend on the formation of a positioned array of nucleosomes, including one containing the 5S RNA gene. In this case nucleosome positioning appears to be not only sequence directed, but also dependent on histone H1 (Chipev and Wolffe, 1992; Fig. 4.10).

In vitro experiments attempt to reconstruct phenomena believed to occur *in vivo*, thereby offering mechanistic insights into a process. However, many of our clearest insights into true gene regulation events in chromosomes come directly from the documentation of *in vivo* events (Wolffe, 1990a). The general observation is that promoters organized into chromatin as seen in the living cell are often accessible to *trans*-acting factors, even when regulatory elements are adjacent to nucleosomes or actually incorporated into them. This conclusion might be contrasted with the vast majority of *in vitro* experiments (Section 4.2).

One of the best examples of *in vivo* control of gene expression in a chromosomal context concerns the regulation of the PHO5 gene of *Saccharomyces cerevisiae*. The yeast *PHO5* gene encodes an acid phosphatase that is induced by a reduction in inorganic phosphate

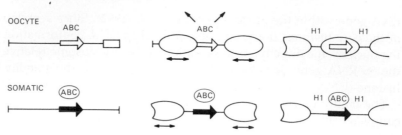

Figure 4.10.   Model for the regulation of *Xenopus* oocyte and somatic 5S RNA genes.
Oocyte (open arrow) and somatic (dark arrow) 5S RNA genes are shown associated with transcription factors (TF)IIIA, B and C forming a transcription complex. Transcription complexes on the oocyte 5S RNA gene are unstable (arrows pointing away from complex), whereas complexes on the somatic 5S RNA gene are stable (circle around complex). In the absence of histone H1, nucleosomes can slide (horizontal double-headed arrows) over the oocyte and somatic 5S RNA genes, especially if transcription factors are not bound to them as in the case of the oocyte 5S RNA genes. When somatic histone H1 accumulates during embryogenesis, nucleosomes are locked into position and the oocyte 5S RNA genes are stably repressed. Since transcription factors are always associated with the somatic 5S RNA genes these genes cannot be incorporated into nucleosomes.

concentration. Two nucleosomes are positioned to either side of an essential promoter element recognized by the *trans*-acting factor PHO4. A second binding site for PHO4 and a site for another *trans*-acting factor PHO2 are incorporated into one of the positioned nucleosomes. On induction, all four positioned nucleosomes are displaced or disrupted. Analysis of mutants reveals that both PHO2 and PHO4 are essential for both the transcriptional activation of the *PHO5* gene and the rearrangement of chromatin structure. These changes are believed to be initiated by the binding of PHO4 to regulatory elements between the two central nucleosomes. Changes in DNA sequence in the adjacent nucleosomes can influence transcriptional activation and chromatin rearrangement, suggesting that the chromosomal organization of the whole promoter region is essential for correct regulation (Almer and Horz, 1986; Almer *et al.*, 1986; Fascher *et al.*, 1990; Straka and Horz, 1991). The precise placement of regulatory elements between or within nucleosomes is clearly important for regulation of this promoter.

An important point is that it is difficult to distinguish between nucleosome displacement, i.e. removal of a complete histone octamer from DNA, and nucleosome disruption, i.e. a conformational change or displacement of an H2A/H2B dimer, on *PHO5* induction (Pavlovic and Horz, 1988). The only certain way is to use protein–DNA cross-linking reagents and to examine the association of specific histones with particular DNA fragments containing the promoter element of interest. Using this technique Varshavsky, Mirzebekov and colleagues have presented evidence for continued histone–DNA contacts in actively transcribed genes (Solomon *et al.*, 1988; Nacheva *et al.*, 1989) (Section 4.3.2). Related to this issue are the experiments of Grunstein and colleagues who have shown through genetic manipulation of histone stoichiometry (Section 5.2), that disruption of the chromatin structure of the *PHO5* promoter even in the absence of induction can significantly activate transcription (Han *et al.*, 1988; Han and Grunstein, 1988). This further establishes that the repression of *PHO5* transcription is related to the chromosomal structure of the gene, and that correct *PHO5* gene regulation requires a chromatin template, as is true for many other yeast genes (Grunstein, 1990).

Although experiments with yeast are useful for establishing the major players in the transcriptional regulation of chromatin templates, significant differences exist between the chromosomal architecture of yeast and vertebrates. The *Saccharomyces cerevisiae* core histones are relatively divergent from those of vertebrates and a protein resembling the histone H1 of vertebrate somatic cells does not appear to exist (Section 2.1.2). Yeast histones are more heavily acetylated than those found in normal vertebrate cells and nucleosomes are relatively tightly packed together. It is therefore important to examine the general applicability of any models established in yeast. Pre-eminent among the systems exploited to this end is the regulation by glucocorticoids of transcription of the mouse mammary tumor virus (MMTV) long terminal repeat (LTR).

Hager and colleagues established that the MMTV LTR is incorporated into six positioned nucleosomes in both episomes and within a mouse chromosome (Fig. 4.11). Induction of transcription by glucocorticoids requires binding of the glucocorticoid receptor (GR) to LTR, disruption of the local chromatin structure, presumably through the GR binding to recognition sequences within nucleosomes, and the assembly of a transcription complex over the TATA box (Archer *et al.*, 1989). Thus comparable events occur on both the PHO5 and MMTV LTR promoters: an inducible transcription factor binds, chromatin structure is rearranged, a transcription complex is assembled and transcription is activated.

Vigorous attempts have been made to reconstruct the transcriptional regulation and concomitant chromatin structural changes of the MMTV LTR *in vitro*. The GR, which is a zinc-finger protein, appears to bind nucleosomal DNA with only a slight reduction in affinity

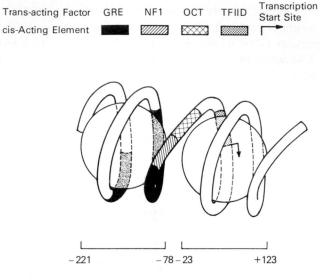

Figure 4.11. The organization of the MMTV LTR into a specific chromatin structure.

Nucleosome positions and key recognition elements for *trans*-acting factors are indicated. The hooked arrow indicates the start site of transcription. Numbers indicate base pair positions of DNA that associates with histones relative to this site.

relative to naked DNA. This interaction may be independent of the precise position of the nucleosome and hence the translational position of the GR binding site in the nucleosome. GR binding is reported to occur when the nucleosome is at −188 to −45 (Pina *et al.*, 1990) and at −219 to −76 (Perlmann and Wrange, 1988) or −221 to −78 (Archer *et al.*, 1991) relative to the start site of transcription (+1). In both instances the rotational orientation of the GR binding site on the surface of the histone octamer will be similar due to the separation of nucleosome boundaries by almost exactly three helical turns of DNA. The latter *in vitro* nucleosome positions compare favorably with those determined *in vivo* (Richard-Foy and Hager, 1987). Surprisingly, association of the GR with the nucleosome containing its binding site, appears to have no effect on the integrity of the structure *in vitro*, unlike the apparent consequence *in vivo*. Binding of the other promoter-specific transcription factors (NF1) which is facilitated by the GR *in vivo*, does not occur *in vitro*. Certain nuclear components that presumably facilitate chromatin structural changes *in vivo* are lacking in the *in vitro* system.

What molecular process causes chromatin structure to be disrupted? The possible ways in which a *cis*-acting element could be incorporated into a nucleosome, and the requisite disruption of the nucleosome required for transcription factor access are shown in Fig. 4.12. The various mechanistic possibilities for disrupting chromatin

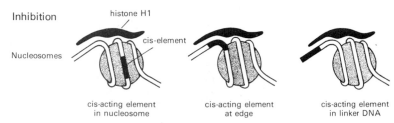

| Inhibition | histone H1 |
| Nucleosomes | cis-element |

| cis-acting element in nucleosome | cis-acting element at edge | cis-acting element in linker DNA |

Figure 4.12. Position of a *cis*-acting element in the nucleosome will influence the contacts of the element with histones.
In the nucleosome, the element will contact the core histones, at the edge it will contact the core histones and histone H1, in the linker DNA it will contact predominantly histone H1.

structure are described in Fig. 4.13. DNA replication is the one event certain to disrupt chromatin and provide access of transcription factors to their cognate sequences (Section 4.3.1). However, DNA replication is not required for chromatin disruption and transcriptional activation of the MMTV LTR (Archer *et al.*, 1989). Schutz and colleagues have extended these studies by examining not only the initial disruption of chromatin structure over the enhancer of the rat tyrosine amino transferase gene (*TAT*) following induction of GR binding, but also the reformation of normal chromatin structure on hormone withdrawal. Both of these changes occur within a few hours, implying that neither disruption or reassembly of nucleosomes is dependent on DNA replication (Reik *et al.*, 1991).

If replication is not involved in chromatin rearrangement, nucleosomes must be disrupted in some alternative way. *Trans*-acting factors might displace histones from nucleosomes directly or might wait for histones to passively exchange out of chromatin before binding to their recognition sequences. In most assays, nucleases are used to examine the incorporation of specific promoter elements into nucleosomes. It is possible that nucleosomes that have altered their position or composition would lose sharp boundaries to nuclease protection even though DNA could remain associated with histones. Protein–DNA cross-linking reagents and antibodies against histones are being used to explore this possibility. Histone H1 and histones H2A/H2B are known to readily exchange in and out of chromatin

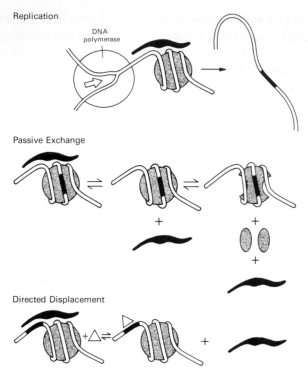

Figure 4.13. Means of disrupting a nucleosome.
Replication may disrupt a nucleosome as DNA duplication requires each
strand of DNA to be a template for second strand synthesis, thus one new
DNA duplex will be naked. It is also possible that DNA polymerase might
displace the histones. Passive exchange requires that an equilibrium
exists between histone bound to DNA and free histones in solution. Such
exchange is known to occur for histone H1 and histones H2A/H2B.
Directed displacement requires that a *trans*-acting factor ($\triangle$) disrupt
histone–DNA contacts.

under physiological conditions (Caron and Thomas, 1981; Louters
and Chakley, 1985) (Sections 3.1.1 and 3.1.2). A certain amount of
DNA folding would occur even in the absence of these proteins. The
capacity to reassemble a complete nucleosome would also remain in
the residual interaction of histones H3/H4 with DNA (Hayes *et al.*,
1991a). Histones H1, H2A and H2B are not able to bind correctly in
chromatin unless histones H3/H4 are already sequestered. It is also
possible that complete histone octamers might exchange from DNA
in vivo. However, this seems unlikely, as solutions of high ionic
strength (> 1 M NaCl) or high temperatures (> 80 °C) are required to
disrupt a nucleosome *in vitro* (Section 2.2.1). Only very basic arginine-
rich proteins, that are unlike known transcription factors, can remove

histones in a natural context (Section 2.5.4). This integrity of the nucleosome is almost entirely due to the stability of the interaction of the arginine-rich histones H3/H4 with DNA (Section 2.2.4). A partial disruption of the nucleosome seems by far the most likely mechanism by which *trans*-acting factors might gain access to DNA in chromatin (Fig. 4.14).

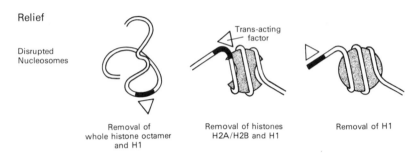

Figure 4.14.   Extent of disruption of nucleosomes shown in Fig. 4.12 required to release the *cis*-acting element depends on where the element is within the nucleosome.

As our information concerning the molecular basis of sequence-directed nucleosome positioning grows (Section 2.2.5) it becomes possible to manipulate specific chromatin structure around the recognition site for a sequence-specific DNA binding protein. Wolffe and Drew made use of synthetic DNA curves and the selective affinity of histone octamers for DNA in this conformation, to manipulate nucleosome position relative to a promoter for bacteriophage T7 RNA polymerase (Wolffe and Drew, 1989). They found that small changes of less than 10 bp in the association of the promoter ($\sim$ 20 bp in length) with the nucleosome could affect the repression of transcription by 10–20-fold. Here even interactions of DNA binding proteins with the periphery of a nucleosome severely inhibits function *in vitro*.

Simpson has carried out similar experiments *in vivo*. Normally in the TRP1ARS1 minichromosome (Section 3.2.3) a nucleosome is not positioned so as to include the ARS1 sequence element required for DNA replication (see Fig. 3.4). However, using a sequence-directed nucleosome-positioning element to move the nucleosome, it was possible to show that incorporation of the 11 bp ARS1 core sequence domain essential for replication into the center of the nucleosomal DNA, but not the peripheral region, inhibited replication of TRP1ARS1 DNA (Simpson, 1990). More experiments of this type will be necessary to prove specific roles for positioned nucleosomes *in vivo*.

Summary

Several studies with positioned nucleosomes have shown that specific *trans*-acting factors can recognize their binding sites when wrapped around the histone core. The histone core appears to be positioned such that the binding sites are accessible to *trans*-acting factors. This suggests that specific chromatin structures assemble in order to fulfill the contrasting needs of DNA compaction and accessibility. Of course, in agreement with experiments using non-specific chromatin structures, certain positioned nucleosomes will inhibit *trans*-acting factor access to DNA. The hypothesis that specific chromatin structures are compatible with nuclear processes gains strength through observations *in vivo*. In yeast and mammalian cells excellent examples exist of promoter elements being regulated through *trans*-acting factors functioning in a chromatin environment and modifying chromatin structure.

4.2.4 *Trans*-acting factors, DNaseI sensitivity, DNaseI-hypersensitive sites and chromosomal architecture

After the early experiments demonstrating the selective association of non-specific DNA binding proteins (prokaryotic RNA polymerases) with transcriptionally active chromatin, Weintraub, Felsenfeld and colleagues were able to show a comparable general accessibility to nucleases (Weintraub and Groudine, 1976; Wood and Felsenfeld, 1982). This general sensitivity to nucleases includes the coding region of a gene and may extend several kilobases to either side of it. DNaseI normally introduces double-strand breaks in this transcriptionally active chromatin over 10 times more frequently than in inactive chromatin. However, the exact structural basis of this generalized sensitivity is unknown. Later we will discuss structural changes in the transcribed regions of genes (Section 4.3.4), however, the dispersed nature of the generalized sensitivity implies other contributory factors. Careful analysis reveals the non-transcribed regions are just as sensitive to DNaseI digestion as are the transcribed regions, provided the last-cut approach to the measurement of DNaseI sensitivity is used. This is defined as digestion by DNaseI to fragment sizes so small that DNA no longer hybridizes to complementary strands after denaturation (Jantzen *et al.*, 1986). A certain length of

DNA is necessary to allow specific recognition (hybridization) of two separated single-stranded regions. An important question that is not yet completely resolved is whether transcription is required to generate generalized nuclease sensitivity in certain instances or whether sensitivity always precedes transcription.

Experiments on the action of mitogens on quiescent cells reveal that a subset of genes, called the immediate-early genes, are rapidly induced (Lau and Nathans, 1987). The most studied examples of such genes are *c-myc* and *c-fos*. These two proto-oncogenes are transcriptionally activated within minutes. Coincident with transcription, the chromatin structure of the proto-oncogenes becomes more accessible to nucleases. Once proto-oncogene transcription ceases, preferential nuclease accessibility is lost (Chen and Allfrey, 1987; Chen *et al.*, 1990; Feng and Villeponteau, 1990). Possible conformational changes in nucleosome structure might account for such effects. It has been proposed that histone H3 sulfhydryl residues might become accessible in nuclease sensitive chromatin; however, transcriptionally active yeast chromatin which has no cysteine and hence no sulfhydryls in the core histones is also retained on the organomercurial agarose columns used to assess nucleosome conformational changes. This suggests that perhaps RNA polymerase or HMGs provide the sulfhydryls that bind to the organomercurial agarose columns retaining active chromatin (Walker *et al.*, 1990). However, the rapidity of the changes in nuclease sensitivity (< 90 s) and their propagation in both directions 5′ and 3′ to the promoter means that transcription and hence RNA polymerase or HMGs cannot account for all of the observed changes consistent with some retention due to histone H3 in higher eukaryotes (Feng and Villeponteau, 1990; Walker *et al.*, 1990; Section 4.3.4). In fact, the speed of the response suggests that changes in nuclease sensitivity precede transcription, and may play a role in regulating *c-fos* expression.

It is interesting that one of the earliest mitogen-induced nuclear signalling events coincident with proto-oncogene induction is the rapid phosphorylation of histone H3 on serine residues within its highly charged, basic N-terminal domain (Section 2.2.4). Whether these changes are localized to chromatin regions containing either *c-fos*, *c-myc* or the other immediate early genes has not yet been determined (Mahadevan *et al.*, 1991). An additional component contributing to the prior sensitization of the proto-oncogenes to nucleases may come from the existence of *trans*-acting factors already associated with the promoter (Herrera *et al.*, 1989). Such interactions are responsible for the second landmark in chromatin: DNaseI hypersensitive sites. These sites are the first place DNaseI introduces a double-strand break in chromatin. They usually involve small segments of DNA sequences (100–200 bp) and are two or more orders of magnitude more accessible to cleavage than in inactive chromatin (Wu *et al.*, 1979; Wu and Gilbert, 1981; McGhee *et al.*, 1981). As the most accessible regions of chromatin to non-histone DNA binding proteins, DNaseI-hypersensitive sites generally denote DNA sequences with important functions in the nucleus.

DNaseI-hypersensitive sites were first detected in the SV40 minichromosome (at the ORI region) and in *Drosophila* chromatin (Elgin, 1988). In general these sites appear to be nucleosome-free and to be accessible to all enzymes or reagents that cut duplex DNA. Higher resolution studies have shown these sites often represent clusters of recognition sites for promoter-specific DNA binding proteins (Emerson *et al.*, 1985). These sites have been mapped to a large number of functional segments of DNA, including promoters, enhancers, locus control elements, transcriptional silencers, origins of replication, recombination elements and structural sites within or around telomeres (Gross and Garrard, 1988).

Around regulated genes these sites often fall into a hierarchy of patterns. In the chicken β-globin locus containing four globin genes (5'ρ-β^H-β^A-ε-3') covering over 65 kb, 12 DNaseI-hypersensitive sites are found (Fig. 4.15). One site is present in all cells independent of

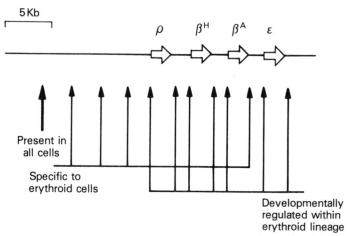

Figure 4.15.   DNaseI-hypersensitive sites of the chicken β-globin locus.

Genes are shown as open arrows; the three different types of DNaseI-hypersensitive sites found are indicated (see Reitman and Felsenfeld, 1990 for details).

whether the genes are transcriptionally active or not. Three sites upstream of the ρ-globin gene were present only in erythroid cells destined to express the globin genes, and have no known functional significance. However, a similar site was found between the β^A and ϵ genes that corresponded to an enhancer element. Four sites were found over the promoters of each gene depending on whether the gene was transcriptionally active, and three sites were found downstream of the genes corresponding to transcription termination elements (the β^A gene excluded). It is important to note that the formation of DNaseI-hypersensitive sites at the promoters of the globin genes is a relatively late step in the commitment of these genes to become transcriptionally active. However, it is clear that the formation of such sites precedes the actual initiation of transcription by RNA polymerase; indeed the generation of these sites may account for a component of the general nuclease sensitivity of a gene (Reitman and Felsenfeld, 1990; Weintraub *et al.*, 1982).

One of the most thorough dissections of a DNaseI-hypersensitive site has been carried out by Elgin and colleagues (Elgin, 1988). The *Drosophila* heat shock protein (*hsp*)26 gene is very rapidly transcriptionally activated by raising the temperature of a fly to a stressful level (a heat shock of 34 °C). Two DNaseI-hypersensitive sites exist at the promoter of the *hsp26* gene, including recognition sequences for the promoter-specific heat shock transcription factor (HSTF, a leucine zipper protein; Section 4.1.6) (Fig. 4.16). Following heat shock HSTF binds to these sites, in contrast TFIID is bound to the TATA box both before and after heat shock (Section 4.1.3). TFIID alone is insufficient to cause the *hsp26* gene to be transcribed, the specific association of the HSTF protein is also required. High-resolution analysis has revealed that a nucleosome is positioned between the proximal and distal binding sites for HSTF, i.e. between the two DNaseI-hypersensitive sites. This nucleosome creates a fixed loop bringing distant HSTF sites close to the TATA box and TFIID. Lis and colleagues have shown that RNA polymerase II is also bound to the promoter next to the TFIID protein. Heat shock and the binding of HSTF allows RNA polymerase II to begin transcriptional elongation by an unknown process (Gilmour and Lis, 1986; Rougvie and Lis, 1988).

Elgin (1988) draws several important conclusions from this analysis. The proximal hypersensitive site exists because there is a region of naked DNA between the specific protein complex at the TATA box and the positioned nucleosome. This region is too small for an additional nucleosome, but contains the binding sites for a promoter-specific factor (HSTF). Thus HSTF can bind to DNA without hindrance from chromatin structure; in fact, the presence of the nucleosome

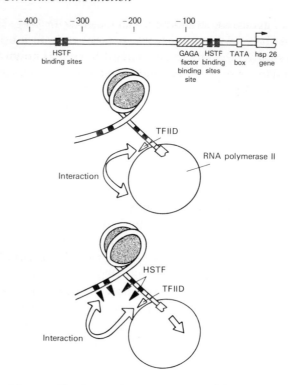

Figure 4.16. The specific chromatin organization of the *Drosophila hsp26* promoter.

Key *cis*-acting elements are indicated relative to the start site of transcription (hooked arrow). The organization of these sites on a specific nucleosomal scaffold is indicated, together with the interactions necessary to prevent or activate transcription. A tethered RNA polymerase II molecule is released through events initiated by the binding of HSTF (heat shock transcription factor).

may create a constrained loop which facilitates interactions between HSTF proteins and TFIID (Section 4.1.1). This is another excellent *in vivo* example of specific proteins regulating gene expression through a specific chromatin structure (Section 4.2.3). This role of chromatin proteins in facilitating a complex regulatory process is similar to the role of HU or IHF in facilitating replication or recombination in prokaryotic systems (Section 4.1.1).

DNaseI-hypersensitive sites have been useful in defining important DNA sequences that have no apparent function (at least when first discovered). Among these were four strongly nuclease-sensitive sites located 10–20 kbp upstream of the cluster of human β-globin genes (Tuan *et al.*, 1985; Grosveld *et al.*, 1987; Forrester *et al.*, 1987). These sites, which have come to be known as locus control regions (LCRs), represent *cis*-acting elements that allow genes that are integrated into a chromosome to be expressed in a way that is independent of chromosomal position, i.e. position effects are abolished (Sections 2.5.5 and 3.2.2). A consequence of this is that LCRs allow each copy of a gene integrated in multiple copies to be expressed equivalently; thus gene expression is copy-number dependent.

When all four DNaseI-hypersensitive sites comprising the LCR are placed adjacent to reporter genes, they appear to function like enhancers. However, three of the four sites do not function as enhancers in transient expression sites, but will only do so after incorporation into the chromosome. This suggests that the LCRs may play a special role in the stabilization of an accessible chromatin structure distinct from the function of a normal enhancer element (Talbot *et al.*, 1990; Philipsen *et al.*, 1990; Talbot and Grosveld, 1991). It is possible that the enhancer-like function of LCRs can be separated from position-independence and copy number dependence, since the region between the chicken β^A and ϵ genes that contains the hypersensitive site acts as an enhancer in transient assays and confers position independence and copy number dependence, but does not confer high levels of gene activity in transgenic mice (Reitman *et al.*, 1990). Alternatively, it could be that chicken sequences do not function well to stimulate transcription in mice because they do not bind the appropriate mammalian promoter-specific proteins.

Each LCR-hypersensitive site contains multiple binding sites for promoter-specific proteins. In the case of the human β-globin genes, one of these sites contains four binding sites for an erythroid tissue sequence-specific DNA binding protein known as GATA-1. In spite of this information, how these sites function is unknown. Several models have been proposed including: unravelling of the chromatin fiber, functions similar to normal enhancers (Section 2.1.2) and stabilization of specific nucleoprotein complexes during replication. Evidence consistent with the propagation of an altered chromatin structure comes from transgenic experiments in which LCRs confer hypersensitive sites and general DNaseI sensitivity whereas promoter elements do not. This suggests that LCRs may be able to overcome the inhibitory influences of localized regions of heterochromatin (Felsenfeld, 1992).

It has been suggested that LCRs or regions adjacent to them might define the boundaries of an active chromatin domain perhaps through attachment to a nuclear matrix or scaffold (Section 2.4.2). Such matrix attachment regions (MARs) or scaffold attachment regions (SARs) have been biochemically defined as those AT-rich

DNA sequences remaining associated with the nuclear matrix or scaffold during its preparation (Section 2.4.2) (Mirkovitch *et al.*, 1984; Cockerill and Garrard, 1986). We have discussed reservations concerning the specificity of such sites as defined *in vitro* (Section 2.4.2). However, a subfamily of sites determined by such procedures (A-elements) can be found at the boundaries of a 24 kbp region of DNaseI-sensitive chromatin containing the chicken lysozyme gene (Phi-Van and Stratling, 1988). Sippel and colleagues have demonstrated that the A-elements insulate a gene from chromosomal effects in stable transformants, but are not required in transient assays for high levels of gene activity (Stief *et al.*, 1989). These A-elements also give high-level position-independent, copy-number-dependent expression of a transgene in transgenic mice (Bonifer *et al.*, 1990). It should be noted that A-elements differ from LCRs in that they function well only when they flank both sides of the gene and its regulatory elements. They also lack intrinsic enhancer activity. The A-elements are good candidates for representing the boundaries of domains of chromatin. How they create these boundaries has not been determined.

Schedl and colleagues have used transposable elements (P-elements) that can introduce stable transformants into the germline of *Drosophila melanogaster*. Using the *hsp70* gene, these investigators defined DNA sequence elements that contained DNaseI-hypersensitive sites to either side of the *hsp70* genes. These specialized chromatin structure (SCS) elements conferred position-independent, copy-number-dependent transcription from the *white* promoter, with the exception of one insertion into heterochromatin within the *Drosophila* X-chromosome. Importantly, the SCS elements do not behave as scaffold attachment regions, establishing a functional separation between the two sequences (Kellum and Schedl, 1991).

An important observation concerning the functions of SCS or A-elements is that they block enhancer function if the element is between the enhancer and the promoter (Eissenberg and Elgin, 1991). Enhancers normally act over very long distances (many kilobases) (Section 4.1.2). It is hard to imagine how a short DNA sequence (100–200 bp) could inhibit DNA looping, suggesting that enhancers in this case are functioning by another mechanism, perhaps related to chromatin folding. Additional insight comes from experiments suggesting that gene regulation can occur through progressively altering the structure of domains of chromatin that are continuous in the chromosome. In the bithorax complex three genes (*Ubx*, *abd-A* and *Abd-B*) are aligned 5' to 3' in the order in which they are expressed during development, and remarkably in the structures whose origin they control, in an anterior to posterior direction in the fly. It has been suggested that these three large (> 20 kb) segments of DNA are each activated in succession (Peifer *et al.*, 1987). Several mutations in the bithorax complex are consistent with this hypothesis since by removing a boundary between two domains of chromatin the anterior–posterior boundary between different structures is also removed (Boulet and Scott, 1988). In this model, boundary elements prevent the propagation of a particular modification of chromatin by functioning as a barrier. One possible means of propagating changes in chromatin or chromosome structure in this way is via the non-histone proteins that are known to repress large domains of chromatin through directing the formation of heterochromatin (Section 2.5.5). Heterochromatin chromatin structure is propagated in *cis* until a boundary is reached (Eissenberg and Elgin, 1991). As discussed above, even an SCS element cannot always prevent this type of chromatin structural transition. In the case of the activation of successive domains of chromatin, it is necessary to postulate a diffusible protein that unravels or activates the chromatin fiber rather than causing its inactivation through formation of heterochromatin.

Summary

Discontinuities in the chromatin fiber are detected as sites that are hypersensitive to DNaseI cleavage. DNaseI-hypersensitive sites can depend on both the binding sites for *trans*-acting factors and also the formation of positioned nucleosomes. They contain histone-free DNA sequences with important regulatory functions in the nucleus. Definition of such sites has led to the discovery of elements that appear to control the function of domains of chromatin–locus control regions, and of elements that appear to define the boundaries between domains of chromatin:A-elements.

DNaseI-hypersensitive sites exist before a gene is transcribed and they may contribute to establishing domains of general sensitivity to nucleases. Such domains may represent unravelled higher order chromatin structures and their formation may be an important prerequisite for transcription.

4.2.5 *Trans*-acting factors and the local organization of chromatin structure

The presence of promoter-specific factors or basal transcription complexes prevents nucleosome assembly on these regions (Section 4.2.2). These non-histone protein–DNA interactions contribute to generating nuclease-hypersensitive sites in the chromatin fiber (Section 4.2.4). Often such hypersensitive sites are found in the midst of ordered arrays of nucleosomes (Almer *et al.*, 1986; Almer and Horz, 1986; Benezra *et al.*, 1986; Richard-Foy and Hager, 1987; Szent-Gyorgi *et al.*, 1987). In these genes the nucleosome arrays exist even within the coding region when transcription is repressed; the arrays are altered or lost when transcription is activated (Section 4.3.2). Similar arrays can be found in centromeres and telomeres (Bloom and Carbon, 1982; Gottschling and Cech, 1984; Budarf and Blackburn, 1986), these arrays contain not hypersensitive sites but regions that are refractory to nuclease cleavage.

Two components clearly contribute to the ordering of nucleosomes relative to nuclease hypersensitive or refractory sites. One is the sequence-directed positioning of nucleosomes (Section 2.2.5), the other is statistical positioning of nucleosomes (Kornberg, 1981), that relies on the generation of boundaries to nucleosome arrays. Statistical positioning depends on nucleosomes packing adjacent to a boundary and then being phased to either side of that boundary due to spacing constraints. The influence of the boundary should decay with distance from the boundaries. Kornberg and colleagues examined nucleosome positioning with respect to the *GAL1–GAL10* intergenic region inserted into *Saccharomyces cerevisiae* minichromosomes. Alterations of the DNA sequences flanking DNaseI-hypersensitive sites left the nucleosomal array unchanged, showing that nucleosome positioning was not a consequence of sequence-directed histone–DNA interactions but depended on proximity to the boundary hypersensitive site. A particular promoter-specific DNA binding protein was found to function as a boundary (Fedor *et al.*, 1988). These experiments are interpreted as representing a passive role of sequence-specific DNA binding proteins in organizing chromatin structure. However, more recent experiments suggest interaction between histones and sequence-specific DNA binding proteins can also occur.

In *S. cerevisiae*, activation of cell-type specific genes depends on expression of the MAT locus (mating type; Dranginis, 1986; Herskowitz, 1989). Two of the proteins regulating gene expression α_2 and α_1, have helix-turn-helix domains (homeodomain proteins; Section 4.1.6). Protein α_2 represses *a*-cell-type specific gene expression through cooperative interactions with another promoter-specific protein, MCM1. The α_2–MCM1 complex associates with a 32 bp sequence located approximately 100 bp upstream of the TATA sequence of five *a*-cell-type specific genes. Repression can also be effected from a comparable distance downstream of the TATA box. In the presence of the α_2–MCM1 complex, nucleosomes are positioned right over the TATA box of at least two of these five genes (Shimizu *et al.*, 1991). The placement of nucleosomes over very different sequences in similar positions suggested that the α_2–MCM1 repressor complex might direct this position. The inclusion of the TATA box into a nucleosome would likely repress transcription (Section 2.1.2).

Interactions between nucleosomal histones and the α_2–MCM1 complex have been suggested by studies with mutants. Although mutations of the N-terminal tails of histones H2A and H2B have little effect on growth or mating ability, deletions of the N-terminal tail of histone H4 leads to a lengthening of the cell cycle and sterility due to aberrant expression of the silent mating loci (Kayne *et al.*, 1988). Activation of these mating loci is specific as a consequence of deletion of the histone H4 tail, since repressed genes such as *PHO5* are not transcribed under these conditions. Individual point mutations of amino acids 16–19 in the N-terminal tail of histone H4 have similar effects to the deletions (Johnson *et al.*, 1990; Megee *et al.*, 1990; Park and Szostak, 1990). One of these point mutations alters a lysine residue that could potentially be acetylated (Section 2.5.2). The effect of the point mutations can be suppressed by mutations in a known regulator of the silent mating loci, SIR3, suggesting direct interaction of the H4 histone tail with *trans*-acting factors (Johnson *et al.*, 1990). Aside from the influence of the histone H4 tail on repression of the silent mating type loci, an intact N-terminus is required for the efficient activation of a number of inducible promoters (Dunin *et al.*, 1991).

Simpson and colleagues have extended these primarily genetic observations to ask direct questions about how a *trans*-acting factor, the α_2–MCM1 repressor, influences chromatin organization. Re-examining the positioning of nucleosomes adjacent to the α_2–MCM1 complex (Shimizu *et al.*, 1991) it was found that deletions of the histone H4 N-terminal tail prevented formation of a stably positioned nucleosome next to the α_2–MCM1 complex (Fig. 4.17). Individual point mutations in the tail region had a comparable effect. An *a*-cell specific promoter was derepressed in the mutant strains, suggesting that the positioning of nucleosomes directed by α_2 is essential for gene regulation (Roth *et al.*, 1992). It appears that in some cases histones and *trans*-acting factors do interact to generate specific chromatin structures *in vivo*. Nucleosome positioning and hence chromatin structure is not a passive filling in of the gaps, but is directed by the appropriate *trans*-acting factor.

Summary
Aside from sequence-directed nucleosome positioning, nucleosomes

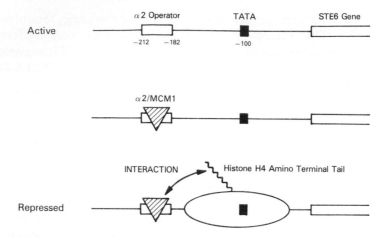

Figure 4.17.   Repression of transcription of the *STE6* gene due to nucleosome positioning directed by the α_2–MCM1 complex (triangle).
This interaction is mediated through the N-terminal tail of histone H4. Base pair positions are indicated relative to the gene.

can be ordered into a specific organization by *trans*-acting factors. This can occur through the generation of boundaries by non-histone proteins that cause nucleosomal arrays to phase themselves relative to these boundaries – statistical positioning. Alternatively, *trans*-acting factors may not passively constrain where nucleosomes form but direct nucleosome positioning through interactions with the core histones. *Trans*-acting factors in some cases have an active role in organizing specific chromatin structures.

4.3 PROCESSIVE ENZYME COMPLEXES AND CHROMATIN STRUCTURE

The compaction of DNA into chromatin provides an obstacle course for any enzyme complex that has to progress along the double helix (Fig. 4.18). DNA and RNA polymerases are large, multi-subunit enzyme complexes that at least transiently unwind the double helix. In eukaryotes these enzymes are over seven times larger than the combined mass of the histones in a nucleosome. The coexistence of a nucleosome and a DNA segment on which a polymerase is attached seems impossible. How these enzymes work in chromatin and the

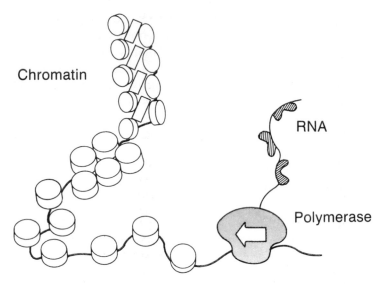

Chromatin

RNA

Polymerase

Figure 4.18. Chromatin structure poses many impediments to the progression of RNA polymerase molecules along the DNA molecule. Presumably the chromatin fiber has to be unwound to allow transcription to occur.

consequences for chromatin structure of their progression through it are important issues.

4.3.1 Replication and the access of transcription factors to DNA

The period in the eukaryotic cell cycle when the genome is duplicated (S-phase) is crucially important for both establishing and maintaining programs of gene activity. The majority of genes in a proliferating cell of a defined type or line continually retain the same states of transcriptional activity through cell division. This reflects the commitment of that cell type to a particular state of determination (Section 4.1.4). How this commitment is established and maintained is not yet resolved; however, several recent experiments suggest a solution to this problem (Wolffe, 1991a). Furthermore, many examples exist in which states of gene activity change following replication events: genes may be either transcriptionally activated or repressed. These events have major consequences for the differentiation of a particular cell type during development. A consideration of the processes occurring at the eukaryotic replication fork again suggests molecular mechanisms that might explain these phenomena.

We have discussed the basic requirements for establishing a eukaryotic gene in a transcriptionally active state (Section 4.1.4). The initial direct binding of transcription factors to DNA is rapid, the sequestration of non-DNA binding factors is relatively slow. *In vitro*, the process of assembling a complete transcription complex takes several minutes. The mutually exclusive binding of the histone octamer or transcription factors to promoter elements in the absence of specific nucleosome positioning is well-established (Section 4.2.2). Prior assembly of nucleosomes prevents transcription factors from binding to DNA and conversely, the prior assembly of a transcription complex prevents nucleosome formation from repressing transcription. Although these results provide an excellent molecular basis for the maintenance of stable states of gene expression in a terminally differentiated non-dividing cell, they do not explain why either transcription complexes or nucleosomes are assembled onto DNA in the first place. Clearly, as both nucleoprotein structures can incorporate the same DNA molecule, the possibility exists of a competition occurring between the assembly of the two structures. As will be discussed later in this section, this competition in fact occurs, however chromatin assembly is staged (Section 3.3). The opportunity of *trans*-acting factors to interact with immature chromatin structures facilitates transcription complex formation. Such a general mechanism for *trans*-acting factor access to DNA could, of course, be assisted by the assembly of specific chromatin structures (Section 4.2.3).

Following replication, nascent DNA is assembled into a chromatin structure that is more sensitive to nucleases than mature chromatin (Section 3.3). The maturation of this nascent chromatin from a nuclease sensitive conformation takes approximately 10–20 min in a mammalian cell. Fractionation of nascent chromatin at various times following replication allowed the demonstration of a two-stage maturation process for chromatin in that histones H3 and H4 are sequestered onto DNA before histones H2A and H2B. Histone H1 is the last histone to be stably incorporated into chromatin. This observation is related to the structure of a nucleosome since histones H3 and H4 form the core of the structure, whereas histones H2A and H2B bind at the periphery of the nucleosome and histone H1 can only associate in its proper place once two turns of DNA are wrapped around the core histones (Section 2.2.4).

The development of *in vitro* replication systems has permitted a further dissection of the assembly of chromatin on replicating DNA (Sections 3.4.2 and 3.4.3). The replication of DNA *in vitro* clearly facilitates chromatin assembly in comparison to non-replicating DNA incubated in the same extracts. The mechanisms underlying this increase in the efficiency of chromatin assembly are unknown, but appear to involve enzymatic processes associated with the replication fork. What is clear is that an intermediate in chromatin assembly forms on replicating DNA, and that this structure has properties consistent with it being a tetramer of histones H3 and H4. The subsequent addition of histones H2A and H2B to this complex generates chromatin with a nuclease sensitivity similar to that expected for the mature state.

In summary, *in vivo* and *in vitro*, chromatin assembly on replicating DNA follows a pathway that might be expected from the structure of the nucleosome. Histones H3 and H4 are deposited first, followed by the sequestration of histones H2A and H2B, and eventually histone H1. This process takes over 20 min to be completed. The nuclease-sensitive stage presumably reflects the accessibility of DNA to the DNA binding nucleases when complexed with only histones H3 and H4. Other DNA binding proteins, such as transcription factors, might also gain access to DNA at this stage in chromatin assembly. Only later when histones H2A, H2B and H1 are sequestered would access to DNA in the nucleosome be restricted.

The presence of a transcription complex (Section 4.1.3) or of a repressive chromatin structure over the promoter elements of a gene would establish a stable state of gene activity in a cell that would not be perturbed until the next round of DNA replication. Which nucleo-protein complex forms on DNA will depend on many variables. For example, the rate of transcription complex formation will principally depend on the concentration of transcription factors and the stability of their interaction with specific DNA sequences (Section 4.1.5). Likewise, nucleosome formation will depend on histone concentrations and the relative stability of a nucleosome containing a particular DNA sequence. Experiments using non-replicating DNA templates for transcription have suggested that the rate of transcription complex formation in the face of chromatin assembly is an important variable in determining gene activity (Section 4.2.2). A second variable follows from the observation that transcription complexes can have different stabilities depending on the promoter elements within the complex. Therefore, the final equilibrium binding of transcription factors during chromatin assembly can also affect gene transcription (Section 4.1.5). These observations do not explain how genes might be effectively programmed in spite of nucleosome assembly; however, they do indicate that transcription factors must be abundant if a transcription complex is to be assembled onto the promoter.

The first insight as to how the efficient programming of genes is accomplished following replication came from experiments using class III genes and non-replicating DNA. The assembly of *Xenopus* 5S RNA genes into complete nucleosomes has been shown to inhibit transcription complex formation (Sections 4.2.2 and 4.2.3). Importantly, the assembly of only histones H3 and H4 onto 5S RNA genes did not inhibit transcription (Tremethick *et al.*, 1990). Similarly, a chromatin template isolated from *Xenopus* sperm, which is naturally deficient in histones H2A, H2B and histone H1 especially after incubation in egg or oocyte extracts, was as active as naked DNA in an *in vitro* transcription reaction (Wolffe, 1989b). These results suggest that a key intermediate in nucleosome assembly, the complex of histones H3 and H4 with DNA, is accessible to transcription factors. This would be consistent with the hypothesis that the newly replicated DNA in the first stage of chromatin maturation could be programmed with transcription complexes.

Xenopus egg extracts have the capacity to carry out complex biological processes such as the complete replication of sperm nuclei (Sections 4.4.2 and 4.4.4). Simplified systems derived from these extracts retain the ability to replicate single-stranded DNA templates with efficiencies of DNA synthesis approaching those observed *in vivo*. The enzymatic processes involved closely resemble those occurring on the lagging strand of the chromosomal replication fork (Section 3.4.2). These extracts also assemble chromatin efficiently on the newly replicated DNA and transcribe class III genes (Wolffe and Brown, 1987; Almouzni and Mechali, 1988a). Using single-stranded DNA as starting material, it was demonstrated that a competition existed between transcription complex assembly on a 5S RNA gene and chromatin assembly on replicating DNA (Almouzni *et al.*, 1990b). This competition depended on the presence of all four core histones. Subsequent work dissected the components of chromatin assembly responsible for the competition between transcription factors and histones for association with the 5S RNA gene. Conditions were established under which the association of histones H2A and H2B with DNA was made limiting (Almouzni *et al.*, 1991). This involved not supplementing the replication reaction with exogenous Mg^{2+}/ATP (Section 3.4.2). Under these circumstances histones H3 and H4 were incorporated into chromatin normally, but complete nucleosomes did not form until either exogenous Mg^{2+}/ATP or exogenous histones H2A and H2B were added. The histone H3/H4 complex was transcriptionally active and apparently accessible to transcription factors. Addition of histones H2A and H2B resulted in the competitive effects on transcription observed in earlier work (Almouzni *et al.*,

1990b, 1991). These competitive effects were shown to be due to the displacement of a non-DNA binding transcription factor (TFIIIB) from DNA, *not* the DNA binding transcription factors themselves. This provides further evidence for the absence of an immediate competition between transcription factors and histones H3 and H4 for binding to the 5S RNA gene. Perhaps the *trans*-acting factors interacting with other regulatory elements with gain access to DNA in a similar manner.

Summary
Experiments using replicating DNA as a substrate demonstrate that a complex of DNA with histones H3 and H4 is initially assembled and that DNA in this complex is accessible to transcription factors (Fig. 4.19). These observations are consistent with experiments on the

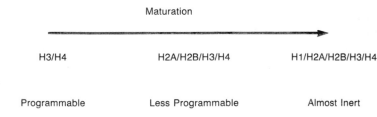

Maturation

H3/H4   H2A/H2B/H3/H4   H1/H2A/H2B/H3/H4

Programmable   Less Programmable   Almost Inert

Figure 4.19. As chromatin structure matures following replication it becomes much less accessible to *trans*-acting factors.

accessibility of chromatin formed on non-replicating DNA to transcription factors. Most importantly they provide confirmation that the molecular mechanisms adopted by the cell to assemble chromatin can accommodate access of essential regulatory elements in DNA to *trans*-acting factors. It is only the subsequent sequestration of histones H2A and H2B followed by histone H1 that render chromatin progressively less accessible to the transcriptional machinery.

4.3.2 The fate of nucleosomes and transcription complexes during replication

The above discussion (Section 4.3.1) has focused on how genes are programmed *de novo*. The experiments we have discussed describe how a gene can be rendered active in spite of ongoing chromatin assembly if transcription factors are available. The presumption implicit in this description is that pre-existing nucleoprotein complexes are disrupted as a consequence of DNA replication. Whether or not this disruption actually occurs is the focus of much current research.

It has long been thought that once established, the nucleoprotein complexes determining gene activity or repression might be capable of reproducing these structures during replication (Tsanev and Sendov, 1971; Brown, 1984; Weintraub, 1985). Under certain conditions, nucleosomes that exist prior to replication can be shown to associate with the daughter DNA molecules (Cusick *et al.*, 1984; Sogo *et al.*, 1986; Bonne-Andrea *et al.*, 1990). However, there is no apparent preference for pre-existing nucleosomes to be reformed on either the leading or lagging strand of the replication fork. This observation allows strong arguments to be made against any imprinting mediated by the arrangement of histones on DNA. This dispersive segregation is consistent with evidence from electron microscopy that histones are displaced from DNA by the replication fork (Sogo *et al.*, 1986). Furthermore, disruption of nucleosomes during replication would explain the presence of both newly synthesized histones and old histones within the same nucleosome in newly assembled chromatin (Jackson, 1987, 1988, 1990). This disruption of chromatin structure during replication suggests a simple mechanism for facilitating gene activation by replication.

As described previously, if transcription factors are not available, the formation of complete nucleosomes following replication will render promoter elements inaccessible unless specific chromatin structures form that mediate access. This repressed state would not be altered by the appearance of transcription factors later in the cell cycle. However, replication events offer the opportunity to disrupt repressive chromatin structures. If transcription factors are present at the instant of replication, the staged assembly of chromatin will facilitate their association with DNA and subsequent gene activation.

Evidence against the dispersive segregation of nucleosomes at the replication fork and thus evidence for the direct templating or reproduction of pre-existing chromatin structures generally came from studies in which histone synthesis was inhibited using inhibitors. Under these conditions replication proceeds, but no new nucleosome assembly occurs (Riley and Weintraub, 1979; Seidman *et al.*, 1979; Handeli *et al.*, 1989). Nucleosomes were observed to segregate to the leading strand replication fork. De Pamphilis and colleagues have recently shown that ongoing protein synthesis is required for DNA synthesis on the lagging strand of the replication fork. Thus duplex DNA, which can be assembled into nucleosomes, is only present on the leading strand (Burhans *et al.*, 1991). Single-stranded DNA which is not assembled efficiently into nucleosomal structures remains on the lagging strand (Almouzni *et al.*, 1990b).

A similar series of experiments has addressed the fate of transcription complexes on DNA during replication (Section 4.1.4). A proposal to explain how transcription complexes might confer a stable state of gene activity through replication came from experiments using *Xenopus* 5S RNA genes (Brown, 1984). Brown proposed the existence of strong cooperative interactions between transcription factors bound to DNA. If just one component of a complex remained associated with a daughter gene after replication, then it could nucleate formation of a complete complex by attracting excess transcription factors from the nucleoplasm. This model was tested by assembling a transcription complex on a 5S RNA gene and replicating through this complex using a viral *in vitro* replication system (Wolffe and Brown, 1986). Replication was dominant to transcription and a direct consequence of replication fork progression through the active 5S RNA gene was the displacement of specific transcription factors. Several correlations from *in vivo* work support this observation. There is a clear antagonism between transcription and replication on efficiently replicating SV40 molecules (Lebkowski *et al.*, 1985; Lewis and Manley, 1985). Replication forks invade the transcriptionally active ribosomal RNA genes in yeast (Saffer and Miller, 1986). The disruption of transcription complexes by replication would provide a simple means of inactivating genes. A replication event through a gene that occurred when transcription factors were not available to bind to the promoter would cause inactivation of the gene through nucleosome assembly unless specific chromatin structures are assembled that might allow access.

A final issue relevant to this discussion is the significance of transcription factors for the initiation of replication and the timing of this initiation in S-phase. If replication disrupts both active and repressed chromatin structures, then the entire nucleus has to be remodeled after each replication event. The accessibility of immature chromatin on newly replicated DNA provides a means to accomplish this remodeling; however, the reformation of nuclear structures has other implications. If there are limiting transcription factors available in a cell then a gene that is replicated early in S-phase has more opportunity to assemble an active transcription complex than a gene that replicates late. This is simply because the gene that replicates early is available for transcription factors to bind before all of the early replicating portion of the genome has sequestered these factors. A late replicating gene will therefore experience a relative deficiency in factor availability (Gottesfeld and Bloomer, 1982; Wormington *et al.*, 1982). Transcriptionally active genes replicate early in S-phase (Goldman *et al.*, 1984; Gilbert, 1986; Guinta and Korn, 1986). The reason for this early replication is unknown, but possibilities include the local disruption of chromatin structure by transcription complexes making that DNA more accessible to the replication machinery (Wolffe and Brown, 1988) and the observation that many transcription factors are, in fact, also replication factors (De Pamphilis, 1988). Although a direct test of the significance of this model has not been made, it remains an attractive mechanism for explaining the maintenance of specific patterns of gene expression in a proliferating cell type.

Our current understanding of the developmental regulation of differential gene expression has followed not only from the definition of *cis*-acting DNA sequence elements and *trans*-acting factors, but also from knowledge of how these function in a chromatin environment in the context of DNA replication events (Wolffe and Brown, 1988). There are many examples that further suggest an integration of these various aspects of nuclear function. A few of these are illustrated below.

The simplest examples of a requirement for replication in the developmental regulation of transcription involve the rapid and precisely programmed life-cycle of eukaryotic viruses. Work with adenoviruses, herpesviruses and vaccinia all indicate that replication events are required for the activation of certain genes. In these instances DNA replication is proposed to remove or dilute inhibitors that deny access of newly synthesized transcription factors to the viral genome (Crossland and Raskas, 1983; Gaynor and Berk, 1983; Thomas and Mathews, 1980; Mavromara-Nazos and Roizman, 1987; Keck *et al.*, 1990). An instance of a replication event being associated with the repression of genes (as opposed to activation) occurs in yeast where a single round of replication inactivates the silent mating-type loci (Miller and Nasmyth, 1984). The replication fork in this case is proposed to remove transcriptional activators.

Early development in both *Drosophila* and *Xenopus* is characterized by rapid rounds of cell division in the absence of transcription. Each of these embryos contain huge stores of nuclear components, including transcription factors and histones (Laskey *et al.*, 1978; Kadonaga, 1990; Tafuri and Wolffe, 1990). DNA replication appears to play a major role in suppressing efficient transcription complex formation, presumably through the repeated disruption of transcription complexes that do form. Procedures that inhibit replication in these embryos lead to the premature activation of transcription (Edgar *et al.*, 1986; Edgar and Schubiger, 1986; Kimelman *et al.*, 1987).

In *Caenorhabditis aenorhabditis elegans* and the sea urchin replication events are correlated with changes in the commitment of cells to a particular developmental fate (Mita-Miyazawa *et al.*, 1985; Edgar and McGhee, 1988). Similar changes can occur in differentiated cells that express one set of specialized genes, and that can only switch to another program of gene expression after one or more cell divisions (e.g. Wolffian regeneration of the lens, Takata *et al.*, 1964). However, replication events are not necessarily essential to change gene expression within a particular cell (Sections 3.1.1 and 3.1.2). This is not surprising since chromatin structure is not completely inert *in vivo*. We have also discussed how the formation of specific chromatin structures can elevate the requirement for DNA replication in order to activate genes (Section 4.2.3).

Summary

DNA replication has an important role in the regulation of eukaryotic gene expression. The replication process transiently disrupts both active and repressed chromatin structures. This provides a key opportunity for mediating changes in the programming of genes. In those cases that have been experimentally tested, imprinting of gene activity mediated by the prior formation of transcription complexes or nucleosomes, followed by their conservative segregation to daughter DNA molecules, has not been observed. However, the mere fact that a gene is active or inactive may influence the time during S-phase when the gene is replicated. By determining the availability of transcription factors at the instant of replication, this may confer continued gene activity or repression.

4.3.3 Chromatin structure and DNA repair

DNA repair presents unusual problems to both *trans*-acting factors and the processive DNA polymerases that must gain access to DNA even in compacted chromatin. DNA damage, most notably pyrimidine dimer formation, has been a particularly useful tool in analyzing the structure of DNA in chromatin (Gale *et al.*, 1987; Gale and Smerdon, 1988; Pehrson, 1989; Section 2.2.3). However, *in vivo* these lesions have to be repaired, through *trans*-acting factors gaining access to DNA in chromatin, excising the lesion and synthesizing new DNA. DNA repair occurs first in regions of DNA not associated with the core histones, and only later within the core DNA (Jensen and Smerdon, 1990). This is in agreement with the general accessibility of DNA binding proteins to linker DNA and core DNA (Section 4.2.1). Nucleosomes have to be rearranged for this second phase of DNA repair to occur. Although increased levels of histone acetylation correlate with the efficiency of DNA repair (Ramanathan and Smerdon, 1989) a more dramatic correlation is with the activity of poly(ADP-ribose) polymerase (Bhatia *et al.*, 1990; Mathis and Althaus, 1990). This enzyme contains zinc-fingers which bind to breaks in the double helix (Gradwohl *et al.*, 1990). Synthesis of long chains of poly(ADP-ribose) has been postulated as a means of competing histones away from nucleosomal structures (Section 2.5.2). Using the yeast minichromosomes, Smerdon and Thoma (1990) examined the rate and efficiency of DNA repair in actively transcribed and inactive regions of chromatin. Transcription of chromatin facilitates DNA repair, suggesting that the passage of RNA polymerase further facilitates access of the repair enzymes to DNA.

Summary
DNA repair enzymes rely on the post-translational modification of histones to locally disrupt chromatin structure. Disruption of chromatin structure by transcription can facilitate this process.

4.3.4 Transcription and chromatin integrity *in vivo*

Some of the most compelling photographs in biology are those made by Miller and colleagues in examining transcription units in *Drosophila* embryos (McKnight *et al.*, 1978; McKnight and Miller, 1979). Active ribosomal RNA genes (class I genes) were found to be densely packed with RNA polymerase I, with few, if any, nucleosomes present. During transcription of ribosomal RNA genes (Conconi *et al.*, 1989), biochemical analysis using psoralen cross-linking (see below) also indicates that nucleosomes disappear from the transcribed sequences. In contrast, non-ribosomal transcription units (class II genes) in which RNA polymerase II molecules were more widely dispersed, were clearly assembled into nucleosomes. However, measurements of the length of DNA within a gene and its compaction into nucleosomal structures revealed that compaction is inversely proportional to the number of RNA polymerase molecules, suggesting that RNA polymerase disrupts nucleosomes at least over the DNA it is bound to.

Most investigators detect nucleosomes on actively transcribed class II genes using nucleases, although the micrococcal nuclease cleavage pattern may become less defined during transcription (Pavlovic *et al.*, 1989). Chemical cross-linking studies also support the continued association of nucleosomes with transcribed class II genes. Formaldehyde cross-linking of histones to DNA in whole cells followed by immunoprecipitation demonstrated a quantitatively similar association of these proteins with transcribed and non-transcribed sequences (Solomon *et al.*, 1988). Slightly different cross-linking methodologies demonstrated that fewer core histone DNA contacts mediated by the globular domains overall were actually established on active genes in comparison to inactive ones. In contrast, contacts by the tails remained unchanged. This suggests that some form of altered chromatin structure exists on transcribed sequences (Karpov *et al.*, 1984; Nacheva *et al.*, 1989). These observations further indicate that the presence of nucleosomes is not incompatible with the efficient elongation by RNA polymerases and that RNA polymerase molecules may displace nucleosomes from DNA. Displacement of nucleosomes could be a consequence of some aspect of the transcription process or simply due to steric occlusion of histone–DNA contacts. At the opposite extreme, Tata and colleagues have shown that the formation of heterochromatin in erythrocyte nuclei (Section 2.5.5) actually inhibits the elongation of engaged RNA polymerase II molecules (Hentschel and Tata, 1978). Clearly the elongation of RNA polymerases can influence chromatin structure and the compaction of chromatin can influence the processivity of DNA polymerases.

Studies on transcriptionally active chromatin have examined the structural basis for its increased accessibility to non-specific DNA binding proteins and nucleases (Section 4.2.1). Among the most noticeable differences between actively transcribed and inactive chromatin is the level of histone acetylation (Csordas, 1990). Nucleosomes released from nuclei at early times of nuclease digestion are enriched in actively transcribed DNA sequences and acetylated histones (Section 2.5.2). More recent studies using organomercurial agarose column chromatography have shown actively transcribed genes to co-purify with highly acetylated histones H4, H3 and H2B (Sterner *et al.*, 1987). Crane-Robinson and colleagues established a more direct linkage between core-histone acetylation and transcribed gene sequences using antibodies against the acetylated lysines of the basic tails to immunoprepitate acetylated chromatin. The acetylated chromatin was found to be enriched in actively transcribed sequences (Hebbes *et al.*, 1988). Although these observations do not establish a causal role for histone acetylation in facilitating elongation through chromatin by RNA polymerase II, they do suggest that acetylation of histones will be at least a consequence of transcribing chromatin.

Evidence consistent with a possible conformational change in the nucleosome is that the histone H3 sulfhydryls of nucleosomes on active genes may be accessible to organomercurial columns, although sulfhydryls of other non-histone proteins might contribute to these results (Chen and Allfrey, 1987; Chen *et al.*, 1990; Section 2.5.2). In transcribed ribosomal RNA genes from *Physarum polycephalum*, the histone H3 sulfhydryls are accessible to chemicals such as iodaceta-mide. This accessibility would imply that the nucleosome would have to be disrupted at least transiently during transcription. Furthermore, electron micrography reveals extended particulate structures on these genes that do not appear to be the same as normal nucleosomes (Prior *et al.*, 1983). It has been proposed that conformational changes in nucleosomes might occur during transcription (Weintraub *et al.*, 1976), although the vast majority of physicochemical studies indicate that a splitting of a nucleosome is very unlikely. It is possible that a dimer of histones H2A and H2B might readily dissociate (Sections 2.2.3 and 2.2.4). This might account for the increase in histone H3 sulfhydryl accessibility.

Nuclease accessibility studies have clearly shown a regular change in a canonical nucleosome repeat in the *Saccharomyces cerevisiae* heat shock protein *(hsp)82* gene. When the promoter of this gene is crip-pled by mutation, transcription ceases and the coding sequence is packaged into a positioned array of nucleosomes. The nucleosomal repeat as seen with DNaseI is approximately 165 bp. When transcrip-tion of the *hsp82* gene is induced by heat shock, the nucleosomal repeat disappears; now DNaseI cuts chromatin approximately every 80–100 bp. It is possible that these new repeats represent an ordered array of non-histone proteins such as RNA polymerases (Lee and Garrard, 1991), although a change in nucleosome conformation is possible. Jackson and colleagues have shown that exchange of his-tones H2A/H2B out of chromatin *in vivo* is facilitated by transcription, whereas little effect is seen for histones H3/H4 (Jackson, 1990). The transcriptionally active fraction of chromatin, enriched in acetylated histones, is deficient in histone H1 and has a particulate structure with a mass consistent with a loss of a dimer of histones H2A/H2B (Locklear *et al.*, 1990). All of these data are consistent with changes in nucleosome structure caused by the transcription process *in vivo*.

Histone acetylation has been shown to correlate with a deficiency of histone H1 in chromatin (Section 2.5.2). Early studies that frac-tionated chromatin based on its solubility at different ionic strengths (Section 2.3.1) suggested that histone H1 was deficient in transcrip-tionally active chromatin (Rose and Garrard, 1984; Rocha *et al.*, 1984; Xu *et al.*, 1986). Different approaches have suggested that histone H1 might still be present in chromatin, albeit interacting differently with DNA. Based on nuclease accessibility studies, Weintraub suggested that the histone H1 in active chromatin could not mediate chromatin folding (Weintraub, 1984). Mirzabekov and colleagues have suggested that in transcribed chromatin, histone H1 no longer associates with DNA through the globular domain, but does so through its basic C- and N-terminal tails (Nacheva *et al.*, 1989). Furthermore, Daneholt and colleagues have used immunoelectron microscopy to show the presence of histone H1 on transcribed Balbiani ring chromatin (Grossbach *et al.*, 1990). UV cross-linking studies suggest that actively transcribed chromatin is slightly deficient in histone H1 and that this deficiency might account for subsequent difficulty in folding the chromatin fiber *in vitro*. Thus, although histone H1 seems to remain in transcribed chromatin to a certain extent, its deficiency might account for the increased acetylation of the core histones, through greater access of the histone acetylase enzyme to its substrate in the chromatin fiber. Alternatively, acetylation of the core histones might impose constraints on the interaction of histone H1 with DNA in chromatin (Section 2.5.2).

Although nucleosome structure may be changed as a consequence of transcription, are nucleosomes actually displaced at the instant of polymerase progression along the DNA in the nucleosome? The evidence discussed earlier suggests that this is indeed the case for class I genes. For RNA polymerase II genes, similar events appear to occur. Psoralen cross-linking has been used to study the consequences for nucleosomes during transcription by RNA polymerase II of SV40 minichromosomes in the living cell. UV light will induce psoralen to cross-link duplex DNA together; however, DNA in the nucleosome reacts less well with psoralen. This is because psoralen has to intercalate into DNA to exert its effects. This is more difficult when DNA is wrapped around the core histones. Thus on denaturation a nucleosome will appear as a single-stranded bubble in the electron microscope. It is also possible to cross-link nascent mRNA to DNA, and to observe such structures adjacent to nucleosome-size single-stranded bubbles. Surprisingly, these results suggest that the RNA polymerase must either coexist with nucleosomes, or that histones must rapidly reassemble to form a nucleosome after RNA polymerase progression through the nucleosome (De Bernardin *et al.*, 1986). These observations are consistent with many electron microscopic studies showing nucleosomes immediately behind an elongating RNA polymerase molecule (McKnight and Miller, 1979; Bjorkroth *et al.*, 1988). Pederson and Morse have examined the topological consequences of transcription in yeast minichromosomes. If nucleosomes unfold or if histones are lost during transcription of a closed circular plasmid *in vivo* a change in supercoiling would be expected. No such changes are observed (Pederson and Morse, 1990). These results demonstrate that nucleosomes must be able to rapidly reform following RNA polymerase II passage.

Summary
RNA polymerases progress through a chromatin template *in vivo*. Nucleosomes appear to be displaced by polymerase, but can rapidly reform after polymerase progression. Histone H1 and a full complement of the core histones are present on transcribed chromatin, but their mode of interaction with DNA may differ from that in the nucleosome. Histone H1 in transcribed regions may be slightly deficient relative to non-transcribed sequences. The core histones are acetylated on transcribed regions with unknown consequences for the processivity of polymerases.

4.3.5 Transcription and chromatin integrity *in vitro*

Although interesting correlations can be made between transcription and chromatin structure *in vivo*, mechanistic studies have to be made *in vitro* in order to understand how RNA polymerase can progress through nucleosomal DNA and what the direct consequences are for the integrity of the nucleosome or chromatin fiber. Felsenfeld and colleagues quantitated both the decrease in accessibility of prokaryotic RNA polymerase to chromatin (Section 4.2.1) and the decrease (66%) in the rate of elongation of the polymerase in chromatin relative to naked DNA (Cedar and Felsenfeld, 1973). Subsequent experiments used a specific template, bacteriophage T7 DNA reconstituted into nucleosomes (Williamson and Felsenfeld, 1978). An important observation was that the rate of elongation by *E. coli* RNA polymerase was reduced at low ionic strength (< 0.1 M) compared to naked DNA, but approached that seen with naked DNA as a template at higher ionic strengths (0.5 M). Nucleosomes do not dissociate at these salt concentration although their interactions with DNA are weakened (Section 3.4.1). These results for the first time suggested that nucleosomes might not present a major impediment to the progress of RNA polymerase in an *in vitro* system.

The next step in resolving how polymerases progress through chromatin came from experiments using defined chromatin templates and purified prokaryotic RNA polymerase. Kornberg and collaborators found that transcription through a single nucleosome by SP6 RNA polymerase occurred without impediment at low ionic strength (40 mM Tris, 6 mM MgCl$_2$). The nucleosome was disrupted by the transcription process (Lorch *et al.*, 1987). An almost identical experiment by Losa and Brown (1987) reached the same conclusion with respect to little impediment existing to SP6 RNA polymerase transcription, but the opposite conclusion with respect to the integrity of the nucleosome. In this case the nucleosome remained intact rather like the 5S RNA gene transcription complex (Section 4.1.4). Subsequent experiments suggested that the explanation for this discrepancy was an unusual stability of the nucleosome positioned on the 5S RNA gene (Lorch *et al.*, 1988). In retrospect this could be due to the intrinsic DNA curvature in 5S DNA (Section 2.2.5). Subsequent experiments with T7 RNA polymerase and a nucleosome assembled on a bacterial sequence revealed that this nucleosome was not displaced by transcription through it (Wolffe and Drew, 1989).

Bradbury and colleagues examined the elongation of T7 RNA polymerase through an array of nucleosomes, finding that nucleosomes remained in place during transcription and that elongation was inhibited over 80% dependent on nucleosome density. Each nucleosome inhibited 15% of RNA polymerases from progressing through it (O'Neill *et al.*, 1992). In contrast to the above experiments Jackson and colleagues found that when nucleosomal templates were transcribed with T7 RNA polymerase in the presence of high levels of topoisomerase I, nucleosomes were disrupted. This demonstrates that under certain conditions nucleosomes cannot reform after transient disruption at the instant of polymerase passage (Pfaffle *et al.*, 1990). Reconciling these disparate experimental results is difficult; however, it is clear that in most circumstances the nucleosome can stay in place or rapidly reform after transcription.

Comparable experiments have been carried out for eukaryotic RNA polymerases. Chambon and colleagues have shown that the rate of transcription elongation for RNA polymerase I and II is greatly decreased for nucleosomal templates. Likewise, several investigators have suggested that RNA polymerase III has difficulty progressing through arrays of nucleosomes (Morse, 1989; Felts *et al.*, 1990). Some of these inhibitory effects could be explained by aggregation of nucleosomal arrays under the ionic conditions used to investigate elongation efficiencies (Hansen and Wolffe, 1992). Free divalent cation (Mg^{2+}) causes the compaction of spaced nucleosomes (Section 2.3.1). Thus the folding of chromatin might further impede RNA polymerase or represent the major constraint to transcription as RNA polymerase II proceeds through a single nucleosome without hindrance (Lorch *et al.*, 1987) whereas multiple nucleosomes cause the polymerase to stall (Izban and Luse, 1991).

How might an RNA polymerase molecule progress through nucleosomal DNA? The nucleosome must be at least transiently disrupted as RNA polymerase progresses through chromatin. This could occur either by the histones individually dissociating from DNA, by the complete octamer of core histones being released and transferred to other DNA regions as a unit, or by the core histones remaining bound but only releasing one or two key contacts at a time to allow the nucleosome to retain its integrity (Thoma, 1991).

Several interesting possibilities follow from the suggestion of Liu and Wang (1987) that processive enzyme complexes, i.e. DNA and RNA polymerases, might transiently introduce positive superhelical stress ahead of the complex as it unwinds duplex DNA, and negative superhelical stress behind it. Although nucleosomes appear stable to both positive and negative superhelical stress (Clark and Felsenfeld, 1991; Clark and Wolffe, 1991), since DNA is overwound in each nucleosome (Section 2.2.3), core histones prefer to associate with DNA that contains negative superhelical turns. It is therefore possible that nucleosome transfer might be favored from the DNA sequences ahead of the RNA polymerase to those behind the enzyme. *In vivo*, it is possible that other cellular components, perhaps those involved in chromatin assembly, might facilitate nucleosome dissolution and reformation. Nucleoplasmin, the *Xenopus* protein responsible for chaperoning histones H2A/H2B in chromatin, is localized to actively transcribed regions (Moreau *et al.*, 1986).

Summary

In vitro experiments indicate that prokaryotic RNA polymerases are not impeded during elongation through the DNA in a single nucleosome. In contrast, arrays of nucleosomes impede both prokaryotic and eukaryotic RNA polymerases. The consequences of transcription are controversial, although it is clear that in some instances the nucleosome remain associated with the template and reforms rapidly once RNA polymerase has elongated through it.

CHAPTER FIVE

Future Prospects

A wealth of opportunities for further progress in understanding how chromatin structure influences nuclear function exist. These occur at several levels ranging through the use of biophysical techniques to the application of molecular genetics.

The detailed information available concerning the structure of the nucleosome allows predictions to be made and experimentally tested concerning the organization of a particular DNA sequence on the histone core. How non-histone proteins recognize DNA after deformation of the double helix in the nucleosome is unknown. Manipulation of the position of specific DNA sequences within the nucleosome will allow this problem to be addressed. Likewise the influence of DNA deformation through association with core or linker histones on the association and dissociation of *trans*-acting factors has not been quantitated. It might be expected that several aspects of gene regulation will be changed by virtue of differences in the recognition of specific chromatin structures rather than of naked DNA by particular proteins.

Modification of the core histones associated with promoter regions have largely unknown consequences for the organization of DNA in the nucleosome and for the subsequent association of non-histone proteins. The effect of such changes on the integrity of the nucleosome is an active area of investigation. Considerable insight into the nucleosome as a structure whose dynamics are regulated by post-translational modification of histones may well be forthcoming.

Although the nucleosome is perhaps the best defined large nucleoprotein complex yet analyzed, the further folding of arrays of nucleosomes is poorly understood. Viewpoints concerning chromatin fiber structure range from a rigid solenoid to the postulate that no such structure exists *in vivo*. Fortunately, systems are now available that allow pure components to be reconstituted into spaced arrays of nucleosomes. Such systems will surely contribute to our understanding of the next step in chromatin assembly – the correct incorporation of linker histones. Assays for this event are now available through the development of techniques to map histone–DNA contacts in terms of DNA and protein sequence. Once again the consequences of this next level of DNA compaction for the access of *trans*-acting factors are only beginning to be defined. The implications of core histone modification, linker histone modification and the incorporation of non-histone structural proteins such as HMGs, for the structure of the chromatin fiber ought to accessible through these studies.

Beyond the chromatin fiber, its further folding through association with the nuclear scaffold is becoming clearer. How this folding is regulated and its significance for nuclear function beyond DNA compaction are subjects of intense interest. The relevance of specific DNA sequences such as boundary elements and locus control regions to establishing domains of chromatin activity is a topic of particular importance. Transgenic mice and *Drosophila* offer a means of defining the DNA sequences involved in mediating domain-specific effects; however, in order to establish the molecular mechanisms responsible for these phenomena such domains of chromatin will have to be reconstructed *in vitro* together with their functional properties.

The capacity to reconstruct chromatin and nuclear structures *in vitro* has progressed enormously. However, our knowledge of chromosome assembly has not been exploited fully. Biochemical dissection of the various assembly extracts should allow the contribution of a particular nuclear architecture to an individual nuclear process to be assessed. Such an approach offers much promise for reconstructing the correct utilization of eukaryotic origins of replication. The staged assembly of chromatin at the replication fork is already seen potentially to explain the access of at least one class of *trans*-acting factors to the DNA regulatory elements of constitutively active genes. The generality of such observations should soon be established.

Many advances in elucidating the specific role of histones in gene regulation have followed from molecular genetic experiments in yeast. The combination of mutational analysis of the yeast histones with functional studies at specific promoters has proven the driving force behind much of the conceptual progress in the field. More elaborate genetic screens are uncovering *trans*-acting factors that appear to interact specifically with both DNA and the histones. Although the final resolution of molecular mechanisms will require the establishment of *in vitro* systems, these genetic approaches are providing a rich store of interesting histone-specific interactions to explore. Although yeast is the '*E. coli* of eukaryotic genetics', and the broad principles of nuclear architecture and function are the same as for higher eukaryotes, sufficient differences exist in the structural proteins of chromatin to make parallel studies using viral episomes extremely important. However there is no doubt that the major players in the field will be defined through work with *Saccharomyces*.

Research on the influence of chromosomal structure on gene expression has undergone a quiet revolution. The long-standing dogma that histones merely package DNA away from any significant role is seen to be as much of an oversimplification as gene regulation occurring on naked DNA. Understanding how both *trans*-acting factors and histones interact to regulate transcription presents molecular biologists with important questions, the answers to which will have general applicability for other processes such as DNA replication, recombination and repair. It is clear that eukaryotic *trans*-acting factors have evolved to operate in a chromatin environment and that histones have evolved to let them function. The future offers considerable promise for reconstructing, and thus understanding, the correct regulation of genes in a natural chromosomal context.

References

Adachi, Y., Kas, E. and Laemmli, U.K. (1989). Preferential, cooperative binding of topoisomerase II to scaffold associated regions. *EMBO J.* **8**, 3997–4006.

Adachi, Y., Luke, M. and Laemmli, U.K. (1991). Chromosome assembly *in vitro*: topoisomerase II is required for condensation. *Cell* **64**, 137–48.

Adhya, S. and Gottesman, M. (1982). Promoter ooclusion: transcription through a promoter may inhibit its activity. *Cell* **29**, 939–44.

Alberts, B., Bray, D., Lewis, J., Raff, M., Roberts, K. and Watson, J.D. (1990). *Molecular Biology of the Cell.* Garland Publishing, London.

Allan, J., Hartman, P.G., Crane-Robinson, C. and Aviles, F.X. (1980). The structure of histone H1 and its location in chromatin. *Nature* **288**, 675–9.

Allan, J., Cowling, G.J., Harborne, N., Cattani, P., Craigie, R. and Gould, H. (1981). Regulation of the higher-order structure of chromatin by histones H1 and H5. *J. Cell Biol.* **90**, 279–88.

Allan, J., Mitchell, T., Harborne, N., Bohm, L. and Crane-Robinson, C. (1986). Roles of H1 domains in determining higher order chromatin structure and H1 location. *J. Mol. Biol.* **187**, 591–601.

Allis, C.D. and Gorovsky, M.A. (1981). Histone phosphorylation in macro- and micronuclei of *Tetrahymena thermophila. Biochemistry* **20**, 3828–33.

Almer, A. and Horz, W. (1986). Nuclease hypersensitive regions with adjacent positioned nucleosomes mark the gene boundaries of the PHO5/PHO3 locus in yeast. *EMBO J.* **5**, 2681–7.

Almer, A., Rudolph, H., Hinnen, A. and Horz, W. (1986). Removal of positioned nucleosomes from the yeast PHO5 promoter upon PHO5 induction releases additional activating DNA elements. *EMBO J.* **5**, 2689–96.

Almouzni, G. and Mechali, M. (1988a). Assembly of spaced chromatin by DNA synthesis in extracts from *Xenopus* eggs. *EMBO J.* **7**, 664–72.

Almouzni, G. and Mechali, M. (1988b). Assembly of spaced chromatin involvement of ATP and DNA topoisomerase activity. *EMBO J.* **7**, 4355–65.

Almouzni, G., Clark, D.J., Mechali, M. and Wolffe, A.P. (1990a). Chromatin assembly on replicating DNA *in vitro. Nucl. Acids Res.* **18**, 5767–74.

Almouzni, G., Mechali, M. and Wolffe, A.P. (1990b). Competition between transcription complex assembly and chromatin assembly on replicating DNA. *EMBO J.* **9**, 573–82.

Almouzni, G., Mechali, M. and Wolffe, A.P. (1991). Transcription complex disruption caused by a transition in chromatin structure. *Molec. Cell. Biol.* **11**, 655–65.

Amati, B.B. and Gasser, S.M. (1988). Chromosomal ARS and CEN elements bind specifically to the yeast nuclear scaffold. *Cell* **54**, 967–78.

Ambrose, C., Rajadhyaksha, A., Lowman, H. and Bina, M. (1989). Locations of nucleosomes on the regulatory region of simian virus 40 chromatin. *J. Molec. Biol.* **209**, 255–63.

Andrews, M.T., Loo, S. and Wilson, L.R. (1991). Coordinate inactivation of class III genes during the gastrula-neurula transition in *Xenopus. Devel. Biol.* **146**, 250–4.

Aparicio, O.M., Billington, B.L. and Gottschling, D.E. (1991). Modifiers of position effect are shared between telomeric and silent mating-type loci in *S. cerevisiae. Cell* **66**, 1279–87.

Archer, T.K., Cordingley, M.G., Marsaud, V., Richard-Foy, H. and Hager, G.L. (1989). Steroid transactivation at a promoter organized in a specifically positioned array of nucleosomes. In *Proceedings: Second International CBT Symposium on the Steroid/Thyroid Receptor Family and Gene Regulation,* Springer Verlag, Berlin, pp. 221–38.

Archer, T.K., Cordingley, M.G., Wolford, R.G. and Hager, G.L. (1991). Transcription factor access is mediated by accurately positioned nucleosomes on the mouse mammary tumor virus promoter. *Molec. Cell. Biol.* **11**, 688–98.

Athey, B.D., Smith, M.F., Rankert, D.A., Williams, S.P. and Langmore, J.P. (1990). The diameters of frozen-hydrated chromatin fibers increase with DNA linker length: evidence in support of variable diameter models for chromatin. *J. Cell Biol.* **111**, 795–806.

Aubert, D., Garcia, M., Benchaibi, M., Poncet, D., Chebloune, Y., Verdier, G., Nigon, V., Samarut, J. and Mura, C.V. (1991). Inhibition of proliferation of primary avian fibroblasts through expression of histone H5 depends on the degree of phosphorylation of the protein. *J. Cell Biol.* **11**, 497–506.

Ausio, J., Dong, F. and van Holde, K.E. (1989). Use of selectively trypsinized nucleosome core particles to analyze the role of the histone tails in the stabilization of the nucleosome. *J. Molec. Biol.* **206**, 451–63.

Avery, O.T., MacLeod, C.M. and McCarty, M. (1944). Studies on the chemical nature of the substance inducing transformation of pneumococcal types. *J. Exp. Med.* **79**, 137–58.

Baer, , B.W. and Rhodes, D. (1983). Eukaryotic RNA polymerase II binds to nucleosome cores from transcribed genes. *Nature (Lond.)* **301**, 482–8.

Banerjee, S. and Cantor, C.R. (1990). Nucleosome assembly of simian virus 40 DNA in a mammalian cell extract. *Molec. Cell Biol.* **10**, 2863–73.

Banerji, J., Olson, L. and Schaffner, W. (1983). A lymphocyte-specific cellular enhancer is located downstream of the joining region in immunoglobulin heavy chain genes. *Cell* **33**, 729–40.

Barry, J.M. and Merriam, R.W. (1972). Swelling of hen erythrocyte nuclei in cytoplasm from *Xenopus* eggs. *Exp. Cell. Res.* **71**, 90–6.

Bateman, E. and Paule, M.R. (1988). Promoter occlusion during ribosomal RNA transcription. *Cell* **54**, 985–92.

Bavykin, S.G., Usachenko, S.I., Zalensky, A.O. and Mirzabekov, A.D. (1990). Structure of nucleosomes and organization of internucleosomal DNA in chromatin. *J. Molec. Biol.* **212**, 495–511.

Becker, P.B., Rabindran, S.K. and Wu, C. (1991). Heat shock-regulated transcription *in vitro* from a reconstituted chromatin template. *Proc. Natl. Acad. Sci. USA* **88**, 4109–13.

Bell, S.P., Learned, R.M., Jantzen, H-M. and Tjian, R. (1988). Functional cooperativity between transcription factors UBF1 and SL1 mediates human ribosomal RNA synthesis. *Science* **241**, 1192–97.

Belmont, A.S., Sedat, J.W. and Agard, D.A. (1987). A three dimensional approach to mitotic chromosome structure: evidence for a complex hierarchical organization. *J. Cell Biol.* **105**, 77–92.

Belmont, A.S., Braunfeld, M.B., Sedat, J.W. and Agard, D.A. (1989). Large scale chromatin structural domains within mitotic and interphase chromosomes *in vivo* and *in vitro*. *Chromosoma* **98**, 129–43.

Benezra, R., Cantor, C.R. and Axel, R. (1986). Nucleosomes are phased along the mouse β-major globin gene in erythroid and non-erythroid cells. *Cell* **44**, 697–704.

Benyajati, C. and Worcel, A. (1976). Isolation, characterization and structure of the folded interphase genome of *Drosophila melanogaster*. *Cell* **9**, 393–407.

Berg, O.G., Winter, R.B. and von Hippel, P.H. (1981). Diffusion-driven mechanisms of protein translocation on nucleic acids. *Biochemistry* **20**, 6929–77.

Berk, A.J. (1986). Adenovirus promoters and E1A transactivation. *Ann. Rev. Genet.* **20**, 45–79.

Berman, H.M. (1991). Hydration of DNA. *Curr. Opin. Struct. Biol.* **1**, 423–7.

Berrios, M. and Avilion, A.A. (1990). Nuclear formation in a *Drosophila* cell-free system. *Exp. Cell Res.* **191**, 64–70.

Bhatia, K., Pommier, Y., Giri, C., Fornace, J., Imaizumi, M., Breitman, T.R., Cherney, B.W. and Smulson, M.E. (1990). Expression of the poly(ADP-ribose) polymerase gene following natural and induced DNA strand breakage and effect of hyperexpression on DNA repair. *Carcinogenesis* **11**, 123–8.

Bienz, M. and Pelham, H.R.B. (1986). Heat shock regulatory elements function as an inclucible enhancer in the *Xenopus* hsp 70 gene and when linked to a heterologous promoter. *Cell* **45**, 753–60.

Bjorkroth, B., Ericsson, C., Lamb, M.M. and Daneholt, B. (1988). Structure of the chromatin axis during transcription. *Chromosoma* **96**, 333–40.

Blasquez, V.C., Xu, M., Moses, S.C. and Garrard, W.T. (1989). Immunoglobulin K gene expression after stable integration. 1. Role of the intronic MAR and enhancer in plasmacytoma cells. *J. Biol. Chem.* **264**, 21183–9.

Blau, H.M. and Baltimore, D. (1991). Differentiation requires continuous regulation. *J. Cell Biol.* **112**, 781–3.

Blau, H.M., Chiu, C-P. and Webster, C. (1983). Cytoplasmic activation of human nuclear genes in stable heterokaryons. *Cell* **32**, 1171–80.

Blau, H.M., Parlath, G.K., Hardeman, E.C., Chiu, C-P, Silberstein, L., Webster, S.F., Miller, S.C. and Webster, C. (1985). Plasticity of the differentiated state. *Science* **230**, 758–66.

Bloom, K.S. and Carbon, J. (1982). Yeast centromere DNA is a unique and highly ordered structure in chromosomes and small circular mini-chromosomes. *Cell* **29**, 305–17.

Blow, J.J. and Laskey, R.A. (1986). Initiation of DNA replication in nuclei and purified DNA by a cell-free extract of *Xenopus* eggs. *Cell* **47**, 577–87.

Blow, J.J. and Sleeman, A.M. (1990). Replication of purified DNA in *Xenopus* egg extract is dependent on nuclear assembly. *J. Cell Sci.* **95**, 383–91.

Bogenhagen, D.F., Wormington, W.M., and Brown, D.D. (1982). Stable transcription complexes of *Xenopus* 5S RNA genes: a means to maintain the differentiated state. *Cell* **28**, 413–21.

Bonifer, C., Vidal, M., Grosveld, F. and Sippel, A.E. (1990). Tissue specific and position independent expression of the complete gene domain for chicken lysozyme in transgenic mice. *EMBO J.* **9**, 2843–8.

Bonne-Andrea, C., Harper, F., Sobczak, J. and De Recondo, A-M (1984). Rat liver HMG1: a physiological nucleosome assembly factor. *EMBO J.* **3**, 1193–9.

Bonne-Andrea, C., Wong, M.L. and Alberts, B.M. (1990). *In vitro* replication through nucleosomes without histone displacement. *Nature* **343**, 719–26.

Bonner, W.M., Wu, R.S., Panusz, H.T. and Muneses, C. (1988). Kinetics of accumulation and depletion of soluble newly synthesized histone is the reciprocal regulation of histone and DNA synthesis. *Biochemistry* **27**, 6542–50.

Boulet, A.M. and Scott, M.P. (1988). Control elements of the P2 promoter of the *Antennapedia* gene. *Genes Devel.* **2**, 1600–14.

Boulikas, T., Wiseman, J.M. and Garrard, W.T. (1980). Points of contact between histone H1 and the histone octamer. *Proc. Natl. Acad. Sci. USA* **77**, 127–31.

Boy de la Tour, E. and Laemmli, U.K. (1988). The metaphase scaffold is helically folded: sister chromatids have predominantly opposite helical handedness. *Cell* **55**, 937–44.

Bradbury, E.M., Inglis, R.J. and Matthews, H.R. (1974). Control of cell division by very lysine rich histone (F1) phosphorylation. *Nature (Lond.)* **247**, 257–61.

Brand, A.H., Breeden, L., Abraham, J., Sternglanz, R. and Nasmyth, K. (1985). Characterization of a 'silencer' in yeast: a DNA sequence with properties opposite to those of a transcriptional enhancer. *Cell* **41**, 41–8.

Brill, S.J. and Sternglanz, R. (1988). Transcription-dependent DNA supercoiling in yeast topoisomerase mutants. *Cell* **54**, 403–11.

Brown, D.D. (1981). Gene expression in eukaryotes. *Science* **211**, 667–74

Brown, D.D. (1984). The role of stable complexes that repress and activate eukaryotic genes. *Cell* **37**, 359–65.

Budarf, M.L. and Blackburn, E.H. (1986). Chromatin structure of the telomeric region and 3'-nontranscribed spacer of *Tetrahymena* ribosomal RNA genes. *J. Biol. Chem.* **261**, 363–69.

Burch, J.B.E. and Weintraub, H. (1983). Temporal order of chromatin structural changes associated with activation of the major chicken vitellogenin gene. *Cell* **33**, 65–76.

Burhans, W.C., Vassilev, L.T., Wu, J., Sogo, J.M., Nallaseth, F.S. and De Pamphilis, M.L. (1991). Emetine allows identification of origins of mammalian DNA replication by imbalanced DNA synthesis, not through conservative nucleosome segregation. *EMBO J.* **10**, 3419–28.

Callan, H.G. (1986). *Lampbrush Chromosomes.* Springer-Verlag, Berlin.

Callan, H.G., Gall, J.G. and Berg, C.A. (1987). The lampbrush chromosomes of *Xenopus laevis*: preparation, identification and distribution of 5S DNA sequences. *Chromosoma* **95**, 236–50.

Camerini-Otero, R.D. and Zasloff, M.A. (1980). Nucleosomal packaging of the thymidine kinase gene of herpes simplex virus transferred into mouse cells: an actively expressed single copy gene. *Proc. Natl. Acad. Sci. USA* **77**, 5079–83.

Camerini-Otero, R.D., Sollner-Webb, B. and Felsenfeld, G. (1976). The organization of histones and DNA in chromatin: evidence for an arginine-rich histone kernel. *Cell* **8**, 333–47.

Campbell, F.E. and Setzer, D.R. (1991). Displacement of *Xenopus* transcription factor IIIA from a 5S rRNA gene by a transcribing RNA polymerase. *Molec. Cell. Biol.* **11**, 3978–86.

Carey, M. (1991). Mechanistic advances in eukaryotic gene activation. *Curr. Opin. Cell Biol.* **3**, 452–60.

Caron, F. and Thomas, J.O. (1981). Exchange of histone H1 between segments of chromatin. *J. Molec. Biol.* **146**, 513–37.

Cary, P.D., Moss, T. and Bradbury, E.M. (1978). High-resolution proton-magnetic-resonance studies of chromatin core particles. *Eur. J. Biochem.* **89**, 475–82.

Cedar, H. and Felsenfeld, G. (1973). Transcription of chromatin *in vitro*. *J. Molec. Biol.* **77**, 237–54.

Cereghini, S. and Yaniv, M. (1984). Assembly of transfected DNA into chromatin: structural changes in the origin-promoter-enhancer region upon replication. *EMBO J.* **3**, 1243–53.

Challberg, M.D. and Kelly, T.J. (1989). Animal virus replication. *Ann. Rev. Biochem.* **58**, 671–717.

Chao, M.V., Gralla, J.D. and Martinson, H.G. (1980a). *Lac* operator nucleosomes I. Repressor binds specifically to operators within the nucleosome core. *Biochemistry* **19**, 3254–60.

Chao, M.V., Martinson, H.G. and Gralla, J.D. (1980b). *Lac* operator nucleosomes can change conformation to strengthen binding by *Lac* repressor. *Biochemistry* **19**, 3260–9.

Chen, G.L., Yang, L., Rowe, T.C. , Halligan, B.D., Tewey, K.M. and Liu, L.F. (1984). Nonintercalative antitumor drugs interfere with the breakage reunion reaction of mammalian DNA topoisomerase II. *J. Biol. Chem.* **259**, 13560–6.

Chen, T.A. and Allfrey, V.G. (1987). Rapid and reversible changes in nucleosome structure accompany the activation, repression and superinduction of the murine proto-oncogenes c-fos and c-myc. *Proc. Natl. Acad. Sci. USA* **84**, 5252–6.

Chen, T.A., Sterner, R., Cozzolino, A. and Allfrey, V.G. (1990). Reversible and irreversible changes in nucleosome structure along the c-fos and c-myc oncogenes following inhibition of transcription. *J. Molec. Biol.* **212**, 481–93.

Cheng, L. and Kelly, T.J. (1989). The transcriptional activator nuclear factor 1 stimulates the replication of SV40 minichromosomes *in vivo* and *in vitro*. *Cell* **59**, 541–51.

Chipev, C.C. and Wolffe, A.P. (1992). The chromosomal organization of *Xenopus laevis* 5S ribosomal RNA genes *in vivo*. *Molec. Cell. Biol.* **12**, 45–55.

Chiu, C.-P. and Blau, H.M. (1984). Reprogramming cell differentiation in the absence of DNA synthesis. *Cell* **37**, 879–87.

Choi, O-R.B. and Engel, J.D. (1988). Developmental regulation of globin gene switching. *Cell* **55**, 17–26.

Churchill, M.E.A. and Travers, A.A. (1991). Protein motifs that recognize structural features of DNA. *Trends Biochem. Sci.* **16**, 92–7.

Ciliberto, G., Castagnoli, L. and Cortese, R. (1983). Transcription by RNA polymerase III. In *Current Topics in Developmental Biology*, Vol. 18. Academic Press, New York, pp. 59–88.

Clark-Adams, C.D., Norris, D., Osley, M.A., Fassler, J.S. and Winston, F. (1988). Changes in histone gene dosage alter transcription in yeast. *Genes Devel.* **2**, 150–9.

Clark, D.J. and Felsenfeld, G. (1991). Formation of nucleosomes on positively supercoiled DNA. *EMBO J.* **10**, 387–95.

Clark, D.J. and Kimura, T. (1990). Electrostatic mechanism of chromatin folding. *J. Molec. Biol.* **211**, 883–96.

Clark, D.J. and Thomas, J.O. (1986). Salt-dependent co-operative interaction of histone H1 with linker DNA. *J. Molec. Biol.* **187**, 569–80.

Clark, D.J. and Wolffe, A.P. (1991). Superhelical stress and nucleosome mediated repression of 5S RNA gene transcription *in vitro*. *EMBO J.* **10**, 3419–28.

Clark, D.J., Hill, C.S., Martin, S.R. and Thomas, J.O. (1988). α-Helix in the carboxy-terminal domains of histones H1 and H5. *EMBO J.* **7**, 69–75.

Clark, R.J. and Felsenfeld, G. (1971). Structure of chromatin. *Nature, New Biol.* **229**, 101–6.

Cockerill, P.N. and Garrard, W.T. (1986). Chromosomal loop anchorage of the Kappa immunoglobulin gene occurs next to the enhancer in a region. *Cell* **44**, 273–82.

Conconi, A., Widmer, R.M., Koller, T. and Sogo, J. M. (1989). Two different chromatin structures coexist in ribosomal RNA genes throughout the cell cycle. *Cell* **57**, 753–61.

Conrad, M.N., Wright, J.H., Wolf, A.J. and Zakian, V.A. (1990). RAP1 protein interacts with yeast telomeres *in vivo*: over production alters telomere structure and decreases chromosome stability. *Cell* **63**, 739–50.

Cook, P.R. (1991). The nucleoskeleton and the topology of replication. *Cell* **66**, 627–37.

Cook, P.R. and Brazell, I.A. (1975). Supercoils in human DNA. *J. Cell Sci.* **19**, 261–79.

Cordingley, M.G., Riegel, A.T. and Hager, G.L. (1987). Steroid dependent interaction of transcription factors with the inducible promoter of mouse mammary tumor virus *in vivo*. *Cell* **48**, 261–70.

Cotten, M. and Chalkley, R. (1987). Purification of a novel, nucleoplasmin-like protein from somatic nuclei. *EMBO J.* **6**, 3945–54.

Courey, A.J., Holtzman, D.A., Jackson, S.P. and Tjian, R. (1989). Synergistic activation by the glutamine-rich domains of human transcription factor Sp1. *Cell* **59**, 827–36.

Cox, L.S. and Laskey, R.A. (1991). DNA replication occurs at discrete sites in pseudonuclei assembled from purified DNA *in vitro*. *Cell* **66**, 271–5.

Cremisi, C. and Yaniv, M. (1980). Sequential assembly of newly synthesized histones on replicating SV40 DNA. *Biochem. Biophys. Res. Commun.* **92**, 1117–23.

Crossland, L.D. and Raskas, H.J. (1983). Identification of adenovirus genes that require template replication for expression. *J. Virol.* **46**, 737–48.

Croston, G.E., Kerrigan, L.A., Liva, L.M., Marshak, D.R. and Kadonaga, J.T. (1991). Sequence-specific antirepression of histone H1-mediated inhibition of basal RNA polymerase II transcription. *Science* **251**, 643–9.

Csordas, A. (1990). On the biological role of histone acetylation. *Biochem. J.* **265**, 23–38.

Cusick, M.E., Lee, K.-S., DePamphilis, M.L. and Wasserman, P.M. (1983). Structure of chromatin at deoxyribonucleic acid replication forks: Nuclease hypersensitivity results from both prenucleosomal deoxyribonucleic acid and an immature chromatin structure. *Biochemistry* **22**, 3873–84.

Cusick, M.E., DePamphilis, M.L. and Wasserman, P.M. (1984). Dispersive segregation of nucleosomes during replication of Simian Virus 40 Chromosomes. *J. Molec. Biol.* **178**, 249–71.

Darby, M.K., Andrews, M.T. and Brown, D.D. (1988). Transcription complexes that program *Xenopus* 5S RNA genes are stable *in vivo*. *Proc. Natl. Acad. Sci. USA* **85**, 5516–20.

Dean, A., Pederson, D.S. and Simpson, R.T. (1989). Isolation of yeast plasmid chromatin. *Methods Enzymol.* **170**, 26–40.

De Bernardin, W., Koller, T. and Sogo, J.M. (1986). Structure of 'in vivo' transcribing chromatin as studied in SV40 minichromosomes. *J. Mol. Biol.* **191**, 469–82.

De Lange, R.J., Farnbrough, D.M., Smith, E.L. and Bonner, J. (1969a). Calf and pea histone IV: the complete amino acid sequence of calf thymus histone IV; presence of *N*-acetyllysine. *J. Biol. Chem.* **244**, 319–34.

De Lange, R.J., Farnbrough, D.M., Smith, E.L. and Bonner, J. (1969b). Calf and pea histone IV: complete amino acid sequence of pea seedling histone IV; comparison with the homologous calf thymus histone. *J. Biol. Chem.* **244**, 5669–79.

DePamphilis, M.L. (1988). Transcriptional elements as components of eukaryotic origins of DNA replication. *Cell* **52**, 635–8.

Di Bernardino, M.A. (1987). Genomic potential of differentiated cells analyzed by nuclear transplantation. *Am. Zool.* **27**, 623–44.

Di Bernardino, W., Koller, T. and Sogo, J.M. (1986). Structure of *in vivo* transcribing chromatin as studied in simian virus 40 minichromosomes. *J. Molec. Biol.* **191**, 469–82.

Dilworth, S.M., Black, S.J. and Laskey, R.A. (1987). Two complexes that contain histones are required for nucleosome assembly *in vitro*: role of nucleoplasmin and N1 in *Xenopus* egg extracts. *Cell* **51**, 1009–18.

Dimitrov, S.I., Russanova, V.R. and Pashev, I.G. (1987). The globular domain of histone H5 is internally located in the 30 nm fiber: an immunochemical study. *EMBO J.* **6**, 2387–92.

DiNardo, S., Voelkel, K. and Sternglanz, R. (1984). DNA topoisomerase II is required for segregation of daughter molecules at the termination of DNA replication. *Proc. Natl. Acad. Sci. USA* **81**, 2616–20.

Dingwall, C., Dilworth, S.M., Black, S.J., Kearsey, S.E., Cox, L.S. and Laskey, R.A. (1987). Nucleoplasmin cDNA reveals polyglutamic acid tracts and a cluster of sequences homologous to putative nuclear localisation signals. *EMBO J.* **6**, 69–74.

Dranginis, A.M. (1986). Regulation of cell type in yeast by the mating type locus. *Trends Biochem. Sci.* **11**, 328–31.

Drew, H.R. (1984). Structural specificities of five commonly-used DNA nucleases. *J. Molec. Biol.* **176**, 535–57.

Drew, H.R. and Travers, A.A. (1985). DNA bending and its relation to nucleosome positioning. *J. Molec. Biol.* **186**, 773–90.

Drew, H.R., McCall, M.J. and Calladine, C.R. (1988). Recent studies of DNA in the crystal. *Ann. Rev. Cell. Biol.* **4**, 1–20.

Dunin, L.K., Mann, R.K., Kayne, P.S. and Grunstein, M. (1991). Yeast histone H4 N-terminal sequence is required for promoter activation *in vivo*. *Cell* **65**, 1023–31.

Dunphy, W.G. and Newport, J.W. (1988). Unraveling of mitotic control mechanisms. *Cell* **55**, 925–8.

Dynan, W.S. and Tjian, R. (1985). Control of eukaryotic messenger RNA synthesis by sequence-specific DNA-binding proteins. *Nature (Lond.)* **316**, 774–8.

Earnshaw, W.C. (1987). Anionic regions in nuclear proteins. *J. Cell Biol.* **105**, 1479–82.

Earnshaw, W.C. (1988). Mitotic chromosome structure. *Bioessays* **9**, 147–50.

Earnshaw, W.C. (1991). Large scale chromosome structure and organization. *Curr. Opin. Struct. Biol.* **1**, 237–44.

Earnshaw, W.C. and Heck, M.M.S. (1985). Localization of topoisomerase II in mitotic chromosomes. *J. Cell Biol.* **100**, 1716–25.

Earnshaw, W.C., Honda, B.M., Laskey, R.A. and Thomas, J.O. (1980). Assembly of nucleosomes: the reaction involving X. *laevis* nucleoplasmin. *Cell* **21**, 373–83.

Earnshaw, W.C., Halligan, B., Cooke, C.A., Heck, M.M.S. and Liu, L.F.

(1985). Topoisomerase II is a structural component of mitotic chromosome scaffolds. *J. Cell Biol.* **100**, 1706–1715.

Echols, H. (1986). Multiple DNA–protein interactions governing high-precision DNA transactions. *Science* **233**, 1050–6.

Echols, H. (1990). Nucleoprotein structures initiating DNA replication, transcription and site-specific recombination. *J. Biol. Chem.* **265**, 14697–700.

Edgar, B.A. and Schubiger, G. (1986). Parameters controlling transcriptional activation during early *Drosophila* development. *Cell* **44**, 871–7.

Edgar, B.A., Kiehle, C.P. and Schubiger, G. (1986). Cell cycle control by the nucleo-cytoplasmic ratio in early *Drosophila* development. *Cell* **44**, 365–72.

Edgar, L.G. and McGhee, J.D. (1988). DNA synthesis and the control of embryonic gene expression in *C. elegans. Cell* **53**, 589–99.

Einck, L. and Bustin, M. (1985). The intracellular distribution and function of the high mobility group chromosomal proteins. *Exp. Cell Res.* **156**, 295–310.

Eissenberg, J.C. and Elgin, S.C.R. (1991). Boundary functions in the control of gene expression. *Trends Genet.* **7**, 335–40.

Elgin, S.C.R. (1988). The formation and function of DNaseI hypersensitive sites in the process of gene activation. *J. Biol. Chem.* **263**, 19259–62.

Elgin, S.C.R. (1990). Chromatin structure and gene activity. *Curr. Opin. Cell. Biol.* **2**, 437–45.

Emerson, B.M., Lewis, C.D. and Felsenfeld, G. (1985). Interaction of specific nuclear factors with the nuclease-hypersensitive region of the chicken adult β-globin gene: nature of the binding domain. *Cell* **41**, 21–30.

Ephrussi, B. (1972). *Hybridisation of Somatic Cells.* Princeton University Press, Princeton, New Jersey.

Erard, M.S., Belenguer, P., Caizergues-Ferrer, M., Pantaloni, A. and Amalric, F. (1988). A major nucleolar protein, nucleolin, induces chromatin decondensation by binding to histone H1. *Eur. J. Biochem.* **175**, 525–30.

Ericsson, C., Mehlin, H., Björkroth, B., Lamb, M.M. and Daneholt, B. (1989). The ultrastructure of upstream and downstream regions of an active Balbiani ring gene. *Cell* **56**, 631–9.

Ericsson, C., Grossbach, U., Björkroth, B. and Daneholt, B. (1990). Presence of histone H1 on an active Balbiani ring gene. *Cell* **60**, 73–83.

Evans, T., Reitman, M. and Felsenfeld, G. (1988). An erythrocyte specific DNA binding factor recognizes a regulatory sequence common to all chicken globin genes. *Proc. Natl. Acad. Sci. USA* **85**, 5976–80.

Falkner, F.G. and Zachau, H.G. (1984). Correct transcription of an immunoglobulin K gene requires an upstream fraguent containing conserved sequence elements. *Nature (Lond.)* **310**, 71–4.

Fascher, K.D., Schmitz, J. and Horz, W. (1990). Role of *trans*-activating proteins in the generation of active chromatin at the PHO 5 promoter in *S. cerevisiae. EMBO J.* **9**, 2523–8.

Fedor, M.J., Lue, N.F. and Kornberg, R.D. (1988). Statistical positioning of nucleosomes by specific protein-binding to an upstream activating sequence in yeast. *J. Molec. Biol.* **204**, 109–27.

Felsenfeld, G. (1992). Chromatin: an essential part of the transcriptional apparatus. *Nature (Lond.)* **355**, 219–24.

Felsenfeld, G. and McGhee, J.D. (1986). Structure of the 30 nm fiber. *Cell* **44**, 375–7.

Felts, S.J., Weil, P.A. and Chalkley, R. (1990). Transcription factor requirements for *in vitro* formation of transcriptionally competent 5S rRNA gene chromatin. *Molec. Cell. Biol.* **10**, 2390–401.

Feng, J. and Villeponteau, B. (1990). Serum stimulation of the c-fos enhancer induces reversible changes in c-fos chromatin structure. *Molec. Cell. Biol.* **10**, 1126–32.

Filipski, J., Leblanc, J., Youdale, T., Sikorska, M. and Walker, P.R. (1990). Periodicity of DNA folding in higher-order chromatin structures. *EMBO J.* **9**, 1319–27.

Finch, J.T., Lutter, L.C., Rhodes, D., Brown, A.S., Rushton, B., Levitt, M. and Klug, A. (1977). Structure of nucleosome core particles of chromatin. *Nature (Lond.)* **269**, 29–36.

FitzGerald, P.C. and Simpson, R.T. (1985). Effects of sequence alterations in a DNA segment containing the 5S rRNA gene from Lythechinus variegatus on positioning of a nucleosome core particle *in vitro. J. Biol. Chem.* **260**, 15318–24.

Fitzsimmons, D.W. and Wolstenholme, G.E.W. (1976). The structure and function of chromatin. Ciba Foundation Symposium 28, 368 pp.

Forbes, D.J., Kirschner, M.W. and Newport, J.W. (1983). Spontaneous formation of nucleus-like structures around bacteriophage DNA microinjected into *Xenopus* eggs. *Cell* **34**, 13–23.

Forrester, W., Takagawa, S., Papayannopoulou, T., Stamatoyannopoulos, G. and Groudine, M. (1987). Evidence for a locus activation region: The formation of developmentally stable hypersensitive sites in globin-expressing hybrids. *Nucl. Acids Res.* **15**, 10159–77.

Fotedar, R. and Roberts, J.M. (1989). Multistep pathway for replication dependent nucleosome assembly. *Proc. Natl. Acad. Sci. USA* **86**, 6459–63.

Franke, W.W. (1987). Nuclear lamins and cytoplasmic intermediate filament proteins: a growing multigene family. *Cell* **48**, 3–4.

Frankel, A.D. and Kim, P.S. (1991). Molecular structure of transcription factors: implications for gene regulation. *Cell* **65**, 717–19.

Fromental, C., Konoo, M., Nomiyama, H., and Chambon, P. (1988). Cooperativity and hierarchical levels of functional organization in the SV40 enhancer. *Cell* **54**, 943–53.

Gale, J.M. and Smerdon, M.J. (1988). Photofootprint of nucleosome core DNA in intact chromatin having different structural states. *J. Molec. Biol.* **204**, 949–58.

Gale, J.M., Nissen, K.A. and Smerdon, M.J. (1987). UV induced formation of pyrimidine dimers in nucleosome core DNA is strongly modulated with a period of 10.3 bases. *Proc. Natl. Acad. Sci. USA* **84**, 6644–8.

Gargiulo, G. and Worcel, A. (1983). Analysis of chromatin assembled in germinal vesicles of *Xenopus* oocytes. *J. Molec. Biol.* **170**, 699–722.

Gargiulo, G., Razvi, F., Ruberti, I., Mohr, I. and Worcel, A. (1985). Chromatin-specific hypersensitive sites are assembled on a Xenopus histone gene injected into *Xenopus* oocytes. *J. Molec. Biol.* **181**, 333–49.

Garner, M.M. and Felsenfeld, G. (1987). Effect of Z-DNA on nucleosome placement. *J. Molec. Biol.* **196**, 581–90.

Gasser, S.M. and Laemmli, U.K. (1986). The organization of chromatin loops: characterization of a scaffold attachment site. *EMBO J.* **5**, 511–18.

Gasser, S.M., Laroche, T., Falquet, J., Boy de la Tour, E. and Laemmli, U.K. (1986). Metaphase chromosome structure involvement of topoisomerase II. *J. Molec. Biol.* **188**, 613–29.

Gaynor, R.B. and Berk, A.J. (1983). *Cis*-acting induction of adenovirus transcription. *Cell* **33**, 683–93.

Georgiev, G.P. (1969). Histones and the control of gene action. *Ann. Rev. Genet.* **3**, 155–180.

Gilbert, D.M. (1986). Temporal order of replication of *Xenopus laevis* 5S ribosomal RNA gnes in somatic cells. *Proc. Natl. Acad. Sci. USA* **83**, 2924–8.

Gilmour, D.S. and Lis, J.T. (1986). RNA polymerase II interacts with the promoter region of the non-induced hsp 70 gene in *Drosophila melanogaster* cells. *Molec. Cell. Biol.* **6**, 3984–9.

Glass, J.R. and Gerace, L. (1990). Lamins A and C bind and assemble at the surface of mitotic chromosomes. *J. Cell Biol.* **111**, 1047–57.

Glikin, G.C., Ruberti, I. and Worcel, A. (1984). Chromatin assembly in *Xenopus* oocytes: *in vitro* studies. *Cell* **37**, 33–41.

Glotov, B.O., Itkes, A.V., Nikolaev, L.G. and Severin, E.S. (1978). Evidence for close proximity between histones H1 and H3 in chromatin of intact nuclei. *FEBS Lett.* **91**, 149–52.

Goldman, M.A., Holmquist, G.P., Gray, M.C., Caston, L.A. and Nog, A. (1984). Replication timing of genes and middle repetitive sequences. *Science* **224**, 686–92.

Gonzalez, P.J. and Palacian, E. (1989). Interaction of RNA polymerase II with structurally altered nucleosomal particles. *J. Biol. Chem.* **264**, 18457–62.

Gorovsky, M.A. (1986). *Molecular Biology of Ciliated Protozoa*. Academic Press, New York.

Gorovsky, M.A., Pleger, G.L., Keevert, J.B. and Johmann, C.A. (1973). Studies on histone fraction F2A1 in macro and micronuclei of *Tetrahymena pyriformis*. *J. Cell. Biol.* **57**, 773–81.

Gottesfeld, J. and Bloomer, L.S. (1982). Assembly of transcriptionally active 5S RNA gene chromatin *in vitro*. *Cell* **28**, 781–91.

Gottschling, D.E. and Cech, T.R. (1984). Chromatin structure of the molecular ends of Oxytricha macronuclear DNA: phased nucleosomes and a telomeric complex. *Cell* **38**, 501–10.

Gradwohl, G., De Murcia, J.M., Molinete, M., Simonin, F., Koken, M., Hoeijmakers, J.H.J. and De Murcia, G. (1990). The second zinc-finger domain of poly(ADP-ribose) polymerase determines specificity for single-stranded breaks in DNA. *Proc. Natl. Acad. Sci. USA* **87**, 2990–4.

Graham, C.F., Arms, K. and Gurdon, J.B. (1966). The induction of DNA synthesis in frog egg cytoplasm. *Devel. Biol.* **14**, 349–81.

Gralla, J.D. (1991). Transcriptional control-lessons from an *E. coli* promoter data base. *Cell* **66**, 415–18.

Green, G.R. and Poccia, D.L. (1985). Phosphorylation of sea urchin sperm H1 and H2B histones precedes chromatin decondensation and H1 exchange during pronuclear formation. *Devel. Biol.* **108**, 235–45.

Gross, D.S. and Garrard, W.T. (1988). Nuclease hypersensitive sites in chromatin. *Ann. Rev. Biochem.* **57**, 159–97.

Grossbach, E.R., Bjorkroth, B. and Daneholt, B. (1990). Presence of histone H1 on an active Balbiani ring gene. *Cell* **60**, 78–83.

Grosschedl, R. and Marx, M. (1988). Stable propagation of the active transcriptional state of an immunoglobulin gene requires continuous enhancer function. *Cell* **55**, 645–54.

Grosveld, F., van Assendelft, G.B., Greaves, D.R. and Kollias, G. (1987). Position independent, high level expression of the human β-globin gene in transgenic mice. *Cell* **51**, 975–85.

Groudine, M. and Conkin, K.F. (1985). Chromatin structure and *de novo* methylation of sperm DNA: implications for activation of the paternal genome. *Science* **228**, 1061–8.

Grunstein, M. (1990). Histone function in transcription. *Ann. Rev. Cell Biol.* **6**, 643–78.

Gruss, C., Gutierrez, C., Burhans, W.C., DePamphilis, M.L., Koller, T. and Sogo, J.M. (1990). Nucleosome assembly in mammalian cell extracts before and after DNA replication. *EMBO J.* **9**, 2911–22.

Guinta, D.E. and Korn, L.J. (1986). Differential order of replication of *Xenopus laevis* 5S RNA genes. *Molec. Cell. Biol.* **6**, 2537–42.

Gurdon, J.B. (1968). Changes in somatic cell nuclei inserted into growing and maturing amphibian oocytes. *J. Embryol. Exp. Morphol.* **20**, 401–14.

Gurdon, J.B. (1974). *The Control of Gene Expression in Animal Development.* Oxford University Press, Oxford.

Gurdon, J.B. (1976). Injected nuclei in frog eggs: fate, enlargement and chromatin dispersal. *J. Embryol. Exp. Morphol.* **36**, 523–40.

Gurdon, J.B. and Brown, D.D. (1965). Cytoplasmic regulation of RNA synthesis and nucleolus formation in developing embryos of *Xenopus laevis*. *J. Molec. Biol.* **12**, 27–35.

Gurdon, J.B., Partington, G.A. and De Robertis, E.M. (1976). Injected nuclei in frog oocytes: RNA synthesis and protein exchange. *J. Embryol. Exp. Morphol.* **36**, 541–53.

Gurdon, J.B., Dingwall, C., Laskey, R.A. and Korn, L.J. (1982). Developmental inactivity of 5S RNA genes persists when chromosomes are cut between genes. *Nature (Lond.)* **299**, 652–3.

Hahn, S., Buratowski, S., Sharp, P.A. and Guarente, L. (1988). Isolation of the gene encoding yeast TATA binding protein TFIID: a gene identical to the SPT15 supressor of Ty element insertions. *Cell* **58**, 1173–81.

Hai T., Horikoshi, M., Roeder, R.G. and Green, M.R. (1988). Analysis of the role of the transcription factor ATF in the assembly of a functional pre-initiation complex. *Cell* **54**, 1043–51.

Han, M. and Grunstein, M. (1988). Nucleosome loss activates yeast down-stream promoters *in vivo*. *Cell* **55**, 1137–45.

Han, M., Chang, M., Kim, U.-J. and Grunstein, M. (1987). Histone H2B repression causes cell-cycle-specific arrest in yeast: effects on chromosomal segregation, replication and transcription. *Cell* **48**, 589–97.

Han, M., Kim, U.-J., Kayne, P. and Grunstein, M. (1988). Depletion of histone H4 and nucleosomes activates the PHO5 gene in *Saccharomyces cerevisiae*. *EMBO J.* **7**, 2221–8.

Handeli, S., Klar, A., Meuth, M. and Cedar, H. (1989). Mapping replication units in animal cells. *Cell* **57**, 909–20.

Hannon, R., Bateman, E., Allan, J., Harborne, N. and Gould, H. (1984). Control of RNA polymerase binding to chromatin by variations in linker histone composition. *J. Molec. Biol.* **180**, 131–49.

Hannon, R., Richards, E.G. and Gould, H. (1986). Facilitated diffusion of a DNA binding protein on chromatin. *EMBO J.* **5**, 3313–19.

Hansen, J.C. and Wolffe, A.P. (1992). The influence of chromatin folding on transcription initiation and elongation by RNA polymerase III. *Biochemistry* (in press).

Hansen, J.C., Ausio, J., Stanik, V.H. and van Holde, K.E. (1989). Homogeneous reconstituted oligonucleosomes, evidence for salt-dependent folding in the absence of histone H1. *Biochemistry* **28**, 9129–36.

Hansen, J.C., van Holde, K.E. and Lohr, D. (1991). The mechanism of nucleosome assembly onto oligomers of the sea urchin 5S DNA positioning sequence. *J. Biol. Chem.* **266**, 4276–82.

Hard, T., Kellenbach, E., Boelens, R., Maler, B.A., Dahlman, K., Freedman, L.P., Carlstedt-Duke, J., Yamamoto, K.R., Gustafsson, J.A. and Kaptein, R. (1990). Solution structure of the glucocorticoid receptor DNA-binding domain. *Science* **249**, 157–60.

Harland, R.M., Weintraub, H. and McKnight, S.L. (1983). Transcription of DNA injected into *Xenopus* oocytes is influenced by template topology. *Nature (Lond.)* **302**, 38–43.

Hawley, D.K. and McClure, W.R. (1982). Mechanism of activation of transcription initiation from the λ Prm promoter. *J. Molec. Biol.* **157**, 493–525.

Hayes, J.J. and Wolffe, A.P. (1992). Histones H2A/H2B inhibit the interaction of TFIIIA with 5S DNA in a nucleosome. *Proc. Natl. Acad. Sci. USA* **89**, 1229–1233.

Hayes, J., Tullius, T.D. and Wolffe, A.P. (1989). A protein-protein interaction is essential for stable complex formation on a 5S RNA gene. *J. Biol. Chem.* **264**, 6009–12.

Hayes, J.J., Tullius, T.D. and Wolffe, A.P. (1990). The structure of DNA in a nucleosome. *Proc. Natl. Acad. Sci. USA* **87**, 7405–9.

Hayes, J.J., Bashkin, J., Tullius, T.D. and Wolffe, A.P. (1991a). The histone core exerts a dominant constraint on the structure of DNA in a nucleosome. *Biochemistry* **30**, 8434–40.

Hayes, J.J., Clark, D.J. and Wolffe, A.P. (1991b). Histone contributions to the structure of DNA in a nucleosome. *Proc. Natl. Acad. Sci. USA* **88**, 6829–33.

Hebbes, T.R., Thome, A.W. and Crane-Robinson, C. (1988). A direct link between core histone acetylation and transcriptionally active chromatin. *EMBO J.* **7**, 1395–402.

Heck, M.M.S. and Earnshaw, W.C. (1986). Topoisomerase II: a specific marker for cell proliferation. *J. Cell Biol.* **103**, 2569–81.

Hentschel, C.C. and Tata, J.R. (1978). Template-engaged and free RNA polymerases during *Xenopus* erythroid cell maturation. *Devel. Biol.* **65**, 496–507.

Herrera, R.E., Shaw, P.E. and Nordheim, A. (1989). Occupation of the c-fos serum response element *in vivo* by a multi-protein complex is unaltered by growth factor induction. *Nature (Lond.)* **340**, 68–70.

Herskowitz, I. (1989). A regulatory hierarchy for cell specialization in yeast. *Nature (Lond.)* **342**, 749–57.

Hewish, D.R. and Burgoyne, L.A. (1973). Chromatin sub-structure: the digestion of chromatin DNA at regularly spaced sites by a nuclear deoxyribonuclease. *Biochem. Biophys. Res. Commun.* **52**, 504–10.

Hill, C.S. and Thomas, J.O. (1990). Core histone–DNA interactions in sea urchin sperm chromatin. The N-terminal tail of H2B interacts with linker DNA. *Eur. J. Biochem.* **187**, 145–53.

Hill, C.S., Rimmer, J.M., Green, B.N., Finch, J.T. and Thomas, J.O. (1991). Histone-DNA interactions and their modulation by phosphorylation of Ser-Pro-X-Lys/Arg-motifs. *EMBO J.* **10**, 1939–48.

Hogan, M.E., Rooney, T.F. and Austin, R.H. (1987). Evidence for kinks in DNA folding in the nucleosome. *Nature (Lond.)* **328**, 554–7.

Hood, L., Kronenberg, M. and Hunkapiller (1985). T. cell antigen receptors and the immunoglobulin supergene family. *Cell* **40**, 225–9.

Horikoshi, M., Carey, M.F., Kakidani, H. and Roeder, R.G. (1988). Mechanism of action of a yeast activator: direct effect of GAL4 derivatives on mammalian TFIID promoter interactions. *Cell* **54**, 665–9.

Horowitz, D.S. and Wang, J.C. (1984). Torsional rigidity of DNA and length dependence of the free energy of DNA supercoiling. *J. Molec. Biol.* **173**, 75–91.

Horowitz, H. and Platt, T. (1982). Regulation of transcription from tandem and convergent promoters. *Nucl. Acids Res.* **10**, 5447–65.

Hsieh, C-H. and Griffith, J.D. (1988). The terminus of SV40 DNA replication and transcription contains a sharp sequence-directed curve. *Cell* **52**, 535–44.

Huletsky, A., De Murcia, G., Muller, S., Hengartner, M., Menard, L., Lamarre, D. and Poirier, G.G. (1989). The effect of poly(ADP-ribosyl)-ation on native and H1–depleted chromatin. A role of poly(ADP-ribosyl)-ation on core nucleosome structure. *J. Biol. Chem.* **264**, 8878–86.

Ichimura, S., Mita, K. and Zama, M. (1982). Essential role of arginine residues in the folding of deoxyribonucleic acid into nucleosome cores. *Biochemistry* **21**, 5329–34.

Izaurralde, E., Kas, E. and Laemmli, U.K. (1989). Highly preferential nucleation of histone H1 assembly on scaffold associated regions. *J. Molec. Biol.* **210**, 573–85.

Izban, M.G. and Luse, D.S. (1991). Transcription on nucleosomal templates by RNA polymerase II *in vitro*: inhibition of elongation with enhancement of sequence-specific pausing. *Genes Devel.* **5**, 683–96.

Jackson, D.A. and Cook, P.R. (1986). Replication occurs at a nucleoskeleton. *EMBO J.* **5**, 1403–11.

Jackson, D.A., Dickinson, P. and Cook, P.R. (1990). The size of chromatin loops in HeLa cells. *EMBO J.* **9**, 567–71.

Jackson, V. (1987). Deposition of newly synthesized histones: new histones H2A and H2B do not deposit in the same nucleosome with new histones H3 and H4. *Biochemistry* **26**, 2315–25.

Jackson, V. (1988). Deposition of newly synthesized histones: hybrid nucleosomes are not tandemly arranged on daughter DNA strands. *Biochemistry* **27**, 2109–20.

Jackson, V. (1990). *In vivo* studies on the dynamics of histone-DNA interaction: evidence for nucleosome dissolution during replication and transcription and a low level of dissolution independent of both. *Biochemistry* **29**, 719–31.

Jackson, V., Shires, A., Tanphaichitr, N. and Chalkley, R. (1976). Modification to histones immediately after synthesis. *J. Molec. Biol.* **104**, 471–83.

Jacobovits, E.B., Bratosin, S. and Aloni, Y. (1980). A nucleosome free region in SV40 minichromosomes. *Nature (Lond.)* **285**, 263–5.

James, T.C. and Elgin, S.C.R. (1986). Identification of a non histone chromosomal protein associated with heterochromatin in *Drosophila melanogaster* and its gene. *Molec. Cell. Biol.* **6**, 3862–72.

Jantzen, H.M., Admon, A., Bell, S.P. and Tjian, R. (1990). Nuclear transcription factor hUBF contains a DNA-binding motif with homology to HMG proteins. *Nature (Lond.)* **344**, 830–6.

Jantzen, K., Fritton, H.P. and Igo-Kemenes, T. (1986). The DNaseI sensitive domain of the chicken lysozyme gene span 24Kb. *Nucl. Acids Res.* **14**, 6085–99.

Jensen, K.A. and Smerdon, M.J. (1990). DNA repair within nucleosome cores of UV-irradiated human cells. *Biochemistry* **29**, 4773–82.

Jerzmanowski, A. and Cole, R.D. (1990). Flanking sequences of *Xenopus* 5S RNA genes determine differential inhibition of transcription by H1 histone *in vitro*. *J. Biol. Chem.* **265**, 10726–32.

Johns, E.W. (1982). *The HMG Chromosomal Proteins*. Academic Press, New York.

Johnson, L.M., Kayne, P.S., Kahn, E.S. and Grunstein, M. (1990). Genetic evidence for an interaction between SIR3 and histone H4 in the repression of silent mating loci in *Saccharomyces cerevisiae*. *Proc. Natl. Acad. Sci. USA* **87**, 6286–90.

Johnson, P. and McKnight, S. (1989). Eucaryotic transcriptional regulatory proteins. *Ann. Rev. Biochem.* **58**, 799–839.

Kadonaga, J.T. (1990). Assembly and disassembly of the *Drosophila* RNA polymerase II complex during transcription. *J. Biol. Chem.* **265**, 2624–31.

Kamakaka, R.T. and Thomas, J.O. (1990). Chromatin structure of transcriptionally competent and repressed genes. *EMBO J.* **9**, 3997–4006.

Karpov, V.L., Preobrazhenskaya, O.V. and Mirzabekov, A.D. (1984). Chromatin structure of hsp 70 genes, activated by heat shock: selective removal of histones from the coding region and their absence from the 5' region. *Cell* **36**, 423–31.

Kassavetis, G.A., Braun, B.R., Nguyen, L.H. and Geiduschek, E.P. (1989). *S. cerevisiae* TFIIIB is the transcription initiation factor proper of RNA polymerase III, while TFIIIA and TFIIIC are assembly factors. *Cell* **60**, 235–45.

Kayne, P.S., Kim, U.-J., Han, M., Mullen, J.R., Yoshizaki, F. and Grunstein, M. (1988). Extremely conserved histone H4 N terminus is dispensable for growth but essential for repressing the silent mating loci in yeast. *Cell* **55**, 27–39.

Keck, J.G., Baldick, Jr. C.J. and Moss, B. (1990). Roie of DNA replication in vaccinia virus gene expression: a naked template is required for transcription of three late *trans*-activator genes. *Cell* **61**, 801–9.

Kelley, D.E., Pollock, B.A., Atchison, M.L. and Perry, R.T. (1988). The coupling between enhancer activity and hypomethylation of k immunoglobulin genes is developmentally regulated. *Molec. Cell Biol.* **8**, 930–7.

Kellum, R. and Schedl, P. (1991). A position effect assay for boundaries of higher order chromosomal domains. *Cell* **64**, 941–50.

Kempnauer, K.-H., Fanning, E., Otto, B. and Knippers, R. (1980). Maturation of newly replicated chromatin of Simian virus 40 and its host cell. *J. Molec. Biol.* **136**, 359–74.

Kimelman, D., Kirschner, M. and Scherson, T. (1987). The events of the midblastula transition in *Xenopus* are regulated by changes in the cell cycle. *Cell* **48**, 399–407.

Kissinger, C.R., Liu, B., Martin-Blanco, E., Kornberg, T.B. and Pabo, C.O. (1990). Crystal structure of an engrailed homeodomain-DNA complex at 2.8 Å resolution. *Cell* **63**, 579–90.

Kleene, K.C. and Flynn, J.F. (1987). Characterization of a cDNA clone encoding the basic protein, TP2, involved in chromatin condensation during spermiogenesis in the mouse. *J. Biol. Chem.* **262**, 17272–7.

Klein, S., Gerster, T., Picard, D., Radbruch, A. and Schaffner, W. (1985). Evidence for transient requirement of the IgH enhancer. *Nucl. Acids Res.* **13**, 8901–12.

Kleinschmidt, J.A. and Seiter, A. (1988). Identification of domains involved in nuclear uptake and histone binding of protein N1 of *Xenopus laevis*. *EMBO J.* **7**, 1605–14.

Kleinschmidt, J.A. and Steinbeisser, H. (1991). DNA dependent phosphorylation of histone H2A.X during nucleosome assembly in *Xenopus laevis* oocytes: involvement of protein phosphorylation in nucleosome spacing. *EMBO J.* **10**, 3043–50.

Kleinschmidt, J.A., Fortkamp, E., Krohne, G., Zentgraf, H. and Franke, W.W. (1985). Co-existence of two different types of soluble histone complexes in nuclei of *Xenopus laevis* oocytes. *J. Biol. Chem.* **260**, 1166–76.

Kleinschmidt, J.A., Dingwall, C., Maier, G. and Franke, W.W. (1986). Molecular characterization of a karyophilic histone-binding protein: cDNA cloning, amino acid sequence and expression of nuclear protein N1/N2 of *Xenopus laevis*. *EMBO J.* **5**, 3547–52.

Kleinschmidt, J.A., Seiter, A. and Zentgraf, H. (1990). Nucleosome assembly *in vitro*: separate histone transfer and synergistic interaction of native histone complexes purified from nuclei of *Xenopus laevis* oocytes. *EMBO J.* **9**, 1309–18.

Klevit, R.E., Hamiot, J.R. and Horvath, S.J. (1990). Solution structure of a zinc finger domain of yeast ADR1. *Proteins* **7**, 215–26.

Klobutcher, L.A., Jahn, C.L. and Prescott, D.M. (1984). Internal sequences are eliminated from genes during macronuclear development in the ciliated protozoan *Oxytricha nova*. *Cell* **36**, 1045–55.

Klug, A. and Lutter, L.C. (1981). The helical periodicity of DNA on the nucleosome. *Nucl. Acids Res.* **9**, 4267–83.

Knezetic, J.A. and Luse, D.S. (1986). The presence of nucleosomes on a DNA template prevents initiation by RNA polymerase II *in vitro*. *Cell* **45**, 95–104.

Knezetic J.A., Jacob, G.A. and Luse, D.S. (1988). Assembly of RNA polymerase II preinitiation complexes before assembly of nucleosomes allows efficient initiation of transcription on nucleosomal templates. *Molec. Cell Biol.* **8**, 3114–21.

Koo, H-S., Wu, H.M. and Crothers, D.M. (1986). DNA bending at adenine thymine tracts. *Nature (Lond.)* **320**, 501–6.

Kornberg, A. (1988). DNA replication. *J. Biol. Chem.* **263**, 1–4.

Kornberg, R. (1974). Chromatin structure: a repeating unit of histones and DNA. *Science* **184**, 868–71.

Kornberg, R.D. (1981). The location of nucleosomes in chromatin: specific or statistical? *Nature (Lond.)* **292**, 579–80.

Kornberg, R. and Thomas, J.O. (1974). Chromatin structure: oligomers of histones. *Science* **184**, 865–8.

Kossel, A. (1928). *The Protamines and Histones*. Longmans, London.

Krohne, G.K. and Franke, W.W. (1980). Immunological identification and localization of the predominant nuclear protein of the amphibian oocyte nucleus. *Proc. Natl. Acad. Sci. USA* **77**, 1034–8.

La Flamme, S., Acuto, S., Markowitz, D., Vick, L., Landschultz, W. and Bank, A. (1987). Expression of chimeric human β- and δ-globin genes during erythroid differentiation. *J. Biol. Chem.* **262**, 4819–26.

Lambert, S.F. and Thomas, J.O. (1986). Lysine-containing DNA-binding regions on the surface of the histone octamer in the nucleosome core particle. *Eur. J. Biochem.* **160**, 191–201.

Laskey, R.A. and Earnshaw, W.C. (1980). Nucleosome assembly. *Nature (Lond.)* **286**, 763–7.

Laskey, R.A., Honda, B.M., Mills, A.D. and Finch, J.T. (1978). Nucleosomes are assembled by an acidic protein which binds histones and transfers them to DNA. *Nature (Lond.)* **275**, 416–20.

Laskey, R.A., Harland, R.M. and Mechali, M. (1983). Induction of chromosome replication during maturation of amphibian oocytes. CIBA. Found Symposium **98**, pp. 25–36.

Lassar, A., Hamer, D.M. and Roeder, R.G. (1985). Stable transcription complex as a class III gene in a minichromosome. *Molec. Cell. Biol.* **5**, 40–5.

Lau, L.F. and Nathans, D. (1987). Expression of a set of growth-related immediate early genes in BALBc/3T3 cells: coordinate regulation with c-fos and c-myc. *Proc. Natl. Acad. Sci. USA* **84**, 1182–9.

Laybourn, P. and Kadonaga, J.T. (1991). Role of nucleosomal cores and histone H1 in the regulation of transcription by RNA polymerase II. *Science* **254**, 238–45.

Lebkowski, J.S., Clancy, S. and Calos, M.P. (1985). Simian virus 40 replication in adenovirus-transformed human cells antagonizes gene expression. *Nature (Lond.)* **317**, 169–71.

Lee, D., Hayes, J.J. and Wolffe, A.P. (1992). Histone acetylation promotes access of transcription factor TFIIIA to 5S RNA genes incorporated into nucleosomes. (submitted).

Lee, M.S. and Garrard, W.T. (1991). Transcription-induced nucleosome 'splitting' an underlying structure for DNaseI sensitive chromatin. *EMBO J.* **10**, 607–15.

Lee, M.S., Gippert, G.P., Soman, K.V., Case, D.A. and Wright, P.E. (1989). Three dimensional solution structure of a single zinc finger DNA-binding domain. *Science* **245**, 635–7.

Leffak, I.M. (1984). Conservative segregation of nucleosome core histones. *Nature (Lond.)* **307**, 82–5.

Leno, G.H. and Laskey, R.A. (1991). The nuclear membrane determines the timing of DNA replication in *Xenopus* egg extracts. *J. Cell Biol.* **112**, 557–66.

Levitt, M. (1978). How many base-pairs per turn does DNA have in solution and in chromatin? Some theoretical calculations. *Proc. Natl. Acad. Sci. USA* **75**, 640–4.

Lewin, B. (1990). *Genes IV*. Oxford University Press, Oxford.

Lewis, C.D. and Laemmli, U.K. (1982). Higher order metaphase chromosome structure: evidence for metalloprotein interactions. *Cell* **29**, 171–81.

Lewis, E.D. and Manley, J.L. (1985). Repression of simian virus 40 early transcription by viral DNA replication in human 293 cells. *Nature (Lond.)* **317**, 172–5.

Li, J.J. and Kelly, T.J. (1984). Simian virus 40 DNA replication *in vitro*. *Proc. Natl. Acad. Sci. USA* **81**, 6973–7.

Lin, R., Cook, R.G. and Allis, C.D. (1991). Proteolytic removal of core histone amino termini and dephosphorylation of histone H1 correlate with the formation of condensed chromatin and transcriptional silencing during *Tetrahymena* macronuclear development. *Genes Devel.* **5**, 1601–10.

Lin, S-Y and Riggs, A.D. (1975). The general affinity of lac repressor for *E. coli* DNA: implications for gene regulation in procaryotes and eucaryotes. *Cell* **4**, 107–11.

Lin, Y.S. and Green, M.R. (1991). Mechanism of action of an acidic transcriptional activator *in vitro*. *Cell* **64**, 971–82.

Liu, L.F. and Wang, J.C. (1987). Supercoiling of the DNA template during transcription. *Proc. Natl. Acad. Sci. USA* **84**, 7024–7.

Lobell, R.B. and Schlief, R.F. (1990). DNA looping and unlooping by Ara C protein. *Science* **250**, 528–32.

Locklear, L., Risdale, J.A., Bazett-Jones, D.P. and Davie, J.R. (1990). Ultrastructure of transcriptionally competent chromatin. *Nucl. Acids Res.* **18**, 7015–24.

Lohka, M.J. and Masui, Y. (1983). Formation *in vitro*, of sperm pronuclei and mitotic chromosomes induced by amphibian ooplasmic components. *Science* **220**, 719–21.

Lohka, M.J. and Masui, Y. (1984). Roles of the cytosol and cytoplasmic particles in nuclear envelope assembly and sperm pronuclear formation in cell-free preparations from amphibian eggs. *J. Cell Biol.* **98**, 1222–30.

Longo, F.J. (1972). An ultrastructural analysis of mitosis and cytokinesis in the sea urchin, *Arbacia punctulata*. *J. Morphol.* **138**, 207–38.

Lorch, Y., La Pointe, J.W. and Kornberg, R.D. (1987). Nucleosomes inhibit the initiation of transcription but allow chain elongation with the displacement of histones. *Cell* **49**, 203–10.

Lorch, Y., La Pointe, J.W. and Kornberg, R.D. (1988). On the displacement of histones from DNA by transcription. *Cell* **55**, 743–4.

Losa, R. and Brown, D.D. (1987). A bacteriophage RNA polymerase transcribes *in vitro* through a nucleosome core without displacing it. *Cell* **50**, 801–8.

Losa, R., Thoma, F. and Koller, T. (1984). Involvement of the globular domain of histone H1 in the higher order structures of chromatin. *J. Molec. Biol.* **175**, 529–51.

Louters, L. and Chalkley, R. (1985). Exchange of histones H1, H2A and H2B *in vivo*. *Biochemistry* **24**, 3080–5.

Lowary, P.T. and Widom, J. (1989). Higher-order structure of *Saccharomyces cerevisiae* chromatin. *Proc. Natl. Acad. Sci. USA* **86**, 8266–70.

Luerssen, H., Hoyer-Fender, S. and Engel, W. (1989). The nucleotide sequence of rat transition protein 2 (TP2) cDNA. *Nucl. Acids Res.* **17**, 3585.

Luke, M. and Bogenhagen, D. (1989). Quantitation of type II topoisomerase in oocytes and eggs of *Xenopus laevis*. *Devel. Biol.* **136**, 459–68.

Lutter, L. (1978). Kinetic analysis of deoxyribonuclease I cleavage sites in the nucleosome core: evidence for a DNA superhelix. *J. Molec. Biol.* **124**, 391–420.

McDowall, A.W., Smith, J.M. and Dubochet, J. (1986). Cryo-electron microscopy of vitrified chromosomes *in situ*. *EMBO J.* **5**, 1395–402.

McGhee, J.D. and Felsenfeld, G. (1979). Reaction of nucleosome DNA with dimethyl sulfate. *Proc. Natl. Acad. Sci. USA* **76**, 2133–7.

McGhee, J.D. and Felsenfeld, G. (1982). Reconstitution of nucleosome core particles containing glucosylated DNA. *J. Molec. Biol.* **158**, 685–98.

McGhee, J.D., Rau, D.C. and Felsenfeld, G. (1980). Orientation of the nucleosome within the higher order structure of chromatin. *Cell* **22**, 87–96.

McGhee, J.D., Wood, W.I., Dolan, M., Engel, J.D. and Felsenfeld, G. (1981). A 200 base pair region at the 5′ end of the chicken adult β-globin gene is accessible to nuclease digestion. *Cell* **27**, 45–55.

McGhee, J.D., Nickol, J.M., Felsenfeld, G. and Rau, D.C. (1983). Higher order structure of chromatin: orientation of nucleosomes within the 30 nm chromatin solenoid is independent of species and spacer length. *Cell* **33**, 831–41.

McKeon, F., Kirschner, M. and Caput, D. (1986). Homologies in both primary and secondary structure between nuclear envelope and intermediate filament proteins. *Nature (Lond.)* **319**, 463–8.

McKnight, S.L. and Miller, O.L. Jr. (1979). Post replicative nonribosomal transcription units in *D. melanogaster* embryos. *Cell* **17**, 551–63.

McKnight, S.L., Bustin, M. and Miller, O.L. Jr. (1978). Electron microscopic analysis of chromosome metabolism in the *Drosophila melanogaster* embryo. *Cold Spring Harbor Symp. Quant. Biol.* **42**, 741–54.

McStay, B., and Reeder, R.H. (1986). A termination site for *Xenopus* RNA polymease I also acts as an element of an adjacent promoter. *Cell* **47**, 913–20.

Mahadevan, L.C., Willis, A.C. and Barrah, M.J. (1991). Rapid histone H3 phosphorylation in response to growth factors, phorbol esters, okadaic acid and protein synthesis inhibitors. *Cell* **65**, 775–83.

Maryanka, D., Cowling, G.J., Allan, J., Fey, S.J., Huvos, P. and Gould, H. (1979). Transcription of globin genes in reticulocyte chromatin. *FEBS Lett.* **105**, 131–6.

Mastrangelo, I.A., Hough, P.V.C., Wilson, V.G., Wall, J.S., Hainfeld, J.F. and Tegtmeyer, P. (1985). Monomers through trimers of large tumor antigen bind in region I and monomers through tetramers bind in region II of simian virus 40 origin of replication DNA as stable structures in solution. *Proc. Natl. Acad. Sci. USA* **82**, 3626–30.

Mastrangelo, I.A., Courey, A.J., Wall, J.S., Jackson, S.P. and Hough, P.V.C. (1991). DNA looping and Sp1 multimer links: a mechanism for transcriptional synergism and enhancement. *Proc. Natl. Acad. Sci. USA* **88**, 5670–4.

Masumoto, H., Masukata, H., Muro, Y., Nozaki, N. and Okazaki, T. (1989). A human centromere antigen (CENP-B) interacts with a short specific sequence in alphoid DNA, a human centromeric satellite. *J. Cell Biol.* **109**, 1963–73.

Mathis, D.J., Oudet, P., Waslyk, B. and Chambon, P. (1978). Effect of histone acetylation on structure and *in vitro* transcription of chromatin. *Nucl. Acids Res.* **5**, 3523–47.

Mathis, G. and Althaus, F.R. (1990). Uncoupling of DNA excision repair and nucleosomal unfolding in poly(ADP-ribose) depeleted mammalian cells. *Carcinogenesis* **11**, 1237–9.

Matsui, T. (1987). Transcription of adenovirus 2 major late and peptide IX genes under conditions of in vitro nucleosome assembly. *Molec. Cell Biol.* **7**, 1401–8.

Mattaj, I., Lienhard, S., Jiricny, J. and De Robertis, E. (1985). An enhancer-like sequence within the *Xenopus* U2 gene promoter facilitates the formation of stable transcription complexes. *Nature (Lond.)* **316**, 163–7.

Mavromara-Nazos, P. and Roizman, B. (1987). Activation of herpes simplex 1 2 genes by viral DNA replication. *Virology* **161**, 593–8.

Mechali, M. and Harland, R.M. (1982). DNA synthesis in a cell-free system from *Xenopus* eggs: priming and elongation on single stranded DNA in vitro. *Cell* **30**, 93–101.

Meeks-Wagner, D. and Hartwell, L.H. (1986). Normal stoichiometry of histone dimer sets is necessary for high fidelity of mitotic chromosome transmission. *Cell* **44**, 53–63.

Meersseman, G., Pennings, S. and Bradbury, E.M. (1991). Chromatosome positioning on assembled long chromatin. Linker histones affect nucleo-some placement on 5S DNA. *J. Molec. Biol.* **220**, 89–100.

Megee, P.C., Morgan, B.A., Mittman, B.A. and Smith, M.M. (1990). Genetic analysis of histone H4: essential role of lysines subject to acetylation. *Science* **247**, 4932–4.

Meisterernst, M., Horikoshi, M. and Roeder, R.G. (1990). Recombinant yeast TFIID, a general transcription factor, mediates activation by the gene specific factor USF in a chromatin assembly assay. *Proc. Natl. Acad. Sci. USA* **87**, 9153–7.

Merriam, R.W. (1969). Movement of cytoplasmic proteins into nuclei induced to enlarge and initiate DNA or RNA synthesis. *J. Cell. Sci.* **5**, 333–49.

Mertz, J.E. (1982). Linear DNA does not form chromatin containing regularly spaced nucleosomes. *Molec. Cell. Biol.* **2**, 1608–18.

Miller, A.M. and Nasmyth, K.A. (1984). Role of DNA replication in the repression of silent mating type loci in yeast. *Nature (Lond.)* **312**, 247–51.

Mills, A.D., Laskey, R.A., Black, P. and De Robertis, E.M. (1980). An acidic protein which assembles nucleosomes *in vitro* is the most abundant pro-tein in *Xenopus* oocyte nuclei. *J. Molec. Biol.* **139**, 561–8.

Mills, A.D., Blow, J.J., White, J.G., Amos, W.B., Wilcock, D. and Laskey, R.A. (1989). Replication occurs at discrete foci spaced throughout nuclei replicating *in vitro*. *J. Cell. Sci.* **94**, 471–7.

Mirkovitch, J., Mirault, M.E. and Laemmli, U. (1984). Organization of the higher-order chromatin loop: specific DNA attachment sites on nuclear scaffold. *Cell* **39**, 223–32.

Mirzabekov, A.D., Shick, V.V., Belyavsky, A.V., Karpov, V.L. and Bavykin, S.G. (1977). The structure of nucleosomes: the arrangement of histones in the DNA grooves and along the DNA chain. *Cold Spring Harbor Symp. Quant. Biol.* **42**, 149–55.

Mirzabekov, A.D., Bavykin, S.G., Karpov, V.L. Preobrazhenskaya, O.V., Elbradise, K.K., Tuneev, V.M., Melinkova, A.F., Goguadze, E.G., Chen-chick, A.A. and Beabealashvili, R.S. (1982). Structure of nucleosomes, chromatin and RNA polymerase promoter complex as revealed by DNA protein cross-linking. *Cold Spring Harbor Symp. Quant. Biol.* **47**, 503–9.

Mirzabekov, A.D., Pruss, D.V. and Elbralidse, K.K. (1990). Chromatin super-structure-dependent cross-linking with DNA of the histone H5 residues Thr1, His25 and His62. *J. Molec. Biol.* **211**, 479–91.

Mita-Miyazawa, I., Ikegami, S. and Satoh, N. (1985). Histospecific acetylcholi-nesterase development in the presumptive muscle cells ioslated from 16-cell-stage ascidian embryos with respect to the number of DNA repli-cations. *J. Embryol. Exp. Morphol.* **87**, 1–12.

Mitchell, P.J. and Tjian, R. (1989). Transcriptional regulation in mammalian cells by sequence-specific DNA binding proteins. *Science* **245**, 371–8.

Moore, G.D., Sinclair, D.A. and Grigliatti, T.A. (1983). Histone gene multi-plicity and position effect variegation in *Drosophila melanogaster*. *Genetics* **105**, 327–44.

Moreau, N., Angelier, N, Bonnanfant-Jais, M-L., Gounon, P. and Kubisz, P. (1986). Association of nucleoplasmin with transcription products as revealed by immunolocalization in the Amphibian oocyte. *J. Cell Biol.* **103**, 683–90.

Morgan, T.H. (1934). *Embryology and Genetics*. Columbia University Press, New York.

Morse, R.H. (1989). Nucleosomes inhibit both transcriptional initiation and elongation by RNA polymerase III *in vitro*. *EMBO J.* **8**, 2343–51.

Morse, R.H. (1992). Change in pattern of histone binding to DNA upon transcriptional activation. *Trends Biochem. Sci.* **17**, 23–6.

Morse, R.H. and Simpson, R.T. (1988). DNA in the nucleosome. *Cell* **54**, 285–8.

Muller-Storm, H., Sogo, J.M. and Schaffner, W. (1989). An enhancer stimulates transcription in *trans* when attached to the promoter via a promoter bridge. *Cell* **58**, 767–77.

Mhrre, C., McCaw, P.S. and Baltimore, D. (1989). A new DNA binding and dimerization motif in immunoglobulin enhancer binding, daughterless, myoD and myc proteins. *Cell* **56**, 777–83.

Nacheva, G.A., Guschin, D.Y., Preobrazhenskaya, O.V., Karpov, V.L., Elbradise, K.K. and Mirzabekov, A.D. (1989). Change in the pattern of histone binding to DNA upon transcription activation. *Cell* **58**, 27–36.

Nakamura, H., Morita, T. and Sato, C. (1986). Structural organization of replicon domains during DNA synthetic phase in the mammalian nucleus. *Exp. Cell Res.* **165**, 291–7.

Nelson, H.C.M., Finch, J.T., Luisi, B.F. and Klug, A. (1987). The structure of an oligo (dA).oligo (dT) tract and its biological implications. *Nature (Lond.)* **330**, 221–6.

Nelson, T., Wiegand, R. and Brutlag, D. (1981). Ribonucleic acid and other polyanions facilitate chromatin assembly *in vitro*. *Biochemistry* **20**, 2594–601.

Newport, J. (1987). Nuclear reconstitution *in vitro*: stages of assembly around protein-free DNA. *Cell* **48**, 205–17.

Newport, J. and Spann, T. (1987). Disassembly of the nucleus in mitotic extracts: membrance vesicularization, lamin disassembly, and chromosome condensation are independent processes. *Cell* **48**, 219–230.

Newport, J., Wilson, K.L. and Dunphy, W.G. (1990). A lamin-independent pathway for nuclear envelope assembly. *J. Cell Biol.* **111**, 2247–59.

Noll, M. (1974a). Subunit structure of chromatin. *Nucl. Acids Res.* **1**, 1573–8.

Noll, M. (1974b). Internal structure of the chromatin subunit. *Nature (Lond.)* **251**, 249–51.

Noll, M. and Kornberg, R.D. (1977). Action of micrococcal nuclease on chromatin and the location of histone H1. *J. Molec. Biol.* **109**, 393–404.

Norton, V.G., Marvin, K.W., Yau, P. and Bradbury, E.M. (1990). Nucleosome linking member change controlled by acetylation of histones H3 and H4. *J. Biol. Chem.* **265**, 19848–52.

Ohaviano, Y. and Gerace, L. (1985). Phosphorylation of the nuclear lamina during interphase and mitosis. *J. Biol. Chem.* **260**, 624–32.

Ohkuma, Y., Horikoshi, M., Roeder, R.G. and Desplan, C. (1990). Engrailed, a homeodomain protein, can repress *in vitro* transcription by competition with the TATA box binding protein transcription factor TFIID. *Proc. Natl. Acad. Sci. USA* **87**, 2289–93.

Ohlenbusch, H.H., Olivera, B.M., Tuan, D. and Davidson, N. (1967). Selective dissociation of histones from calf thymus nucleoprotein. *J. Molec. Biol.* **25**, 299–315.

Olins, A.L. and Olins, D.E. (1974). Spheroid chromatin units (*v*-bodies). *Science* **183**, 330–2.

Oliva, R., Bazett-Jones, D.P., Locklear, L. and Dixon, G.H. (1990). Histone hyperacetylation can induce unfolding of the nucleosome core particle. *Nucl. Acids Res.* **18**, 2739–47.

O'Neill, T.E., Roberge, M. and Bradbury, E.M. (1992). Nucleosome arrays inhibit both initiation and elongation of transcription by T7 RNA polymerase. *J. Molec. Biol.* **223**, 67–78.

Onnuki, Y. (1968). Structure of chromosomes: morphological studies on the spiral structure of human somatic chromosomes. *Chromosoma* **25**, 402–8.

Ostrowski, M.C., Richard-Foy, H., Wolford, R.G., Berard, D.S. and Hager, G.L. (1983). Glucocorticoid regulation of transcription at an amplified episomal promoter. *Molec. Cell Biol.* **3**, 2045–57.

Otting, G., Qian, Y.Q., Billeter, M., Müller, M., Affolter, M., Gehring, W.J. and Wüthrich, K. (1990). Protein-DNA contacts in the structure of a homeodomain-DNA complex determined by nuclear magnetic resonance spectroscopy in solution. *EMBO J.* **9**, 3085–92.

Pardon, J.F., Worcester, D.L., Wooley, J.C., Tatchell, K., van Holde, K.E. and Richards, B.M. (1975). Low-angle neutron scattering from chromatin subunit particles. *Nucl. Acids Res.* **2**, 2163–75.

Pardue, M.L. and Hennig, W. (1990). Heterochomatin: junk or collectors item? *Chromosoma* **100**, 3–7.

Park, E.C. and Szostak, J.W. (1990). Point mutations in the yeast histone H4 gene prevent silencing of the silent mating type locus HML. *Molec. Cell. Biol.* **10**, 4932–4.

Parker, C.S. and Roeder, R.G. (1977). Selective and accurate transcription of the *Xenopus laevis* 5S RNA genes in isolated chromatin by purified RNA polymerase III. *Proc. Natl. Acad. Sci. USA* **74**, 44–8.

Paule, M.R. (1990). In search of the single factor. *Nature (Lond.)* **344**, 819–20.

Pavletich, N.P. and Pabo, C.O. (1991). Zinc finger-DNA recognition: crystal structure of a Zif 268–DNA complex at 2.1 Å. *Science* **252**, 809–16.

Pavlovic, B. and Horz, W. (1988). The chromatin structure at the promoter of a glyceraldehyde phosphate dehydrogenase gene from *Saccharomyces cerevisiae* reflects its functional state. *Molec. Cell. Biol.* **8**, 5513–20.

Pavlovic, J., Banz, E. and Parish, R.W. (1989). The effects of transcription on the nucleosome structure of four *Dictyostelium* genes. *Nucl. Acids Res.* **17**, 2315–32.

Pays, E., Donaldson, D. and Gilmour, R.S. (1979). Specificity of chromatin transcription *in vitro* anomalies due to RNA-dependent RNA synthesis. *Biochim. Biophys. Acta* **562**, 112–30.

Pederson, D.S. and Morse, R.H. (1990). Effect of transcription of yeast chromatin on DNA topology *in vivo*. *EMBO J.* **9**, 1873–81.

Pederson, D.S., Venkatesan, M., Thoma, F. and Simpson, R.T. (1986). Isolation of an episomal yeast gene and replication origin as chromatin. *Proc. Natl. Acad. Sci. USA* **83**, 7206–10.

Pehrson, J.R. (1989). Thymine dimer formation as a probe of the path of DNA in and between nucleosomes in intact chromatin. *Proc. Natl. Acad. Sci. USA* **86**, 9149–53.

Peifer, M., Karch, F. and Bender, W. (1987). The bithorax complex: control of segment identity. *Genes Devel.* **1**, 891–8.

Pennings, S., Meersseman, G. and Bradbury, E.M. (1991). Mobility of positioned nucleosomes on 5S rDNA. *J. Molec. Biol.* **220**, 101–10.

Perlmann, T. and Wrange, O. (1988). Specific glucocorticoid receptor binding to DNA reconstituted in a nucleosome. *EMBO J.* **7**, 3073–83.

Perry, C.A. and Annunziato, A.T. (1989). Influence of histone acetylation on the solubility, H1 content and DNaseI sensitivity of newly assembled chromatin. *Nucl. Acids. Res.* **17**, 4275–91.

Perry, M. and Chalkley, R. (1981). The effect of histone hyperacetylation on the nuclease sensitivity and the solubility of chromatin. *J. Biol. Chem.* **256**, 3313–18.

Pfaffle, P., Gerlach, V., Bunzel, L. and Jackson, V. (1990). *In vitro* evidence that transcription induced stress causes nucleosome dissolution and regeneration. *J. Biol. Chem.* **265**, 16830–40.

Philipsen, S., Talbot, O., Fraser, P. and Grosveld, F. (1990). The β-globin dominant control region: hypersensitive site. *EMBO J.* **9**, 2159–68.

Philpott, A., Leno, G.H. and Laskey, R.A. (1991). Sperm decondensation in *Xenopus* egg cytoplasm is mediated by nucleoplasmin. *Cell* **65**, 569–78.

Phi-Van, L. and Stratling, W.H. (1988). The matrix attachment regions of the chicken lysozyme gene co-map with the boundaries of the chromatin domain. *EMBO J.* **7**, 655–64.

Pieler, T., Hamm, J. and Roeder, R.G. (1987). The 5S internal control region is composed of three distinct sequence elements, organized as two functional domains with variable spacing. *Cell* **48**, 91–100.

Pina, B., Bruggemeier, U. and Beato, M. (1990). Nucleosome positioning modulates accessibility of regulatory proteins to the mouse mammary tumor virus promoter. *Cell* **60**, 719–31.

Pluta, A.F., Cooke, C.A. and Earnshaw, W.C. (1990). Structure of the human centromere at metaphase. *Trends Biochem. Sci.* **15**, 181–5.

Poccia, D. (1986). Remodeling of nucleoproteins during gametogenesis, fertilization, and early development. *Int. Rev. Cytol.* **105**, 1–65.

Poljak, L.G. and Gralla, J.D. (1987). Competition for formation of nucleosomes on fragmented SV40 DNA: a hyperstable nucleosome forms on the termination region. *Biochemistry* **26**, 295–303.

Prior, C.P., Cantor, C.R., Johnson, E.M., Littau, V.C. and Allfrey, V.G. (1983). Reversible changes in nucleosome structure and histone H3 accessibility in transcriptionally active and inactive states of rDNA chromatin. *Cell* **34**, 1033–942.

Privé, G.G., Yanagi, K. and Dickerson, R.E. (1991). Structure of the B-DNA decamer CCAACGTTGG and comparison with isomorphous decamers CCAAGATTGG and CCAGGCCTGG. *J. Molec. Biol.* **217**, 177–91.

Prunell, A. (1982). Nucleosome reconstitution on plasmid-inserted poly(dA). poly(dT). *EMBO J.* **1**, 173–9.

Ptashne, M. (1986). Gene regulation by proteins acting nearly and at a distance. *Nature (Lond.)* **322**, 697–701.

Pugh, B.F. and Tjian, R. (1990). Mechanism of transcriptional activation by SP1: evidence for coactivators. *Cell* **61**, 1187–97.

Pugh, B.F. and Tjian, R. (1992). Diverse transcriptional functions of the multi-subunit eukaryotic TFIID complex. *J. Biol. Chem.* **267**, 679–82.

Ramanathan, B. and Smerdon, M.J. (1989). Enhanced DNA repair synthesis in hyperacetylated nucleosomes. *J. Biol. Chem.* **264**, 11026–34.

Ramsay, N., Felsenfeld, G., Rushton, B.M. and McGhee, J.D. (1984). A 145 base pair DNA sequence that positions itself precisely and asymmetrically on the nucleosome core. *EMBO J.* **3**, 2605–11.

Rasmussen, R., Benvegnu, D., O'Shea, E.K., Kim, P.S. and Albe, T. (1991). X-ray scattering indicates that the leucine zipper is a coiled coil. *Proc. Natl. Acad. Sci. USA* **88**, 561–4.

Rattner, J.B. and Lin, C.C. (1985). Radical loops and helical coils coexist in metaphase chromosomes. *Cell* **42**, 291–6.

Reeves, R., Gorman, C.M. and Howard, B. (1985). Minichromosome assembly of non-integrated plasmid DNA transfected into mammalian cells. *Nucl. Acids Res.* **13**, 3599–615.

Reik, A., Schutz, G. and Stewart, A.F. (1991). Glucocorticoids are required for establishment and maintenance of an alteration in chromation structure: induction leads to a reversible disruption of nucleosomes over an enhancer. *EMBO J.* **10**, 2569–76.

Reitman, M. and Felsenfeld, G. (1990). Developmental regulation of topoisomerase II sites and DNaseI-hypersensitive sites in the chicken β-globin locus. *Molec. Cell. Biol.* **10**, 2774–86.

Reitman, M., Lee, E., Westphal, H. and Felsenfeld, G. (1990). Site-independent expression of the chicken A-globin gene in transgenic mice. *Nature (Lon.)* **348**, 749–752.

Reuter, G., Giarre, M., Farah, J., Gausz, J., Spierer, A. and Spierer, P. (1990). Dependence of position-effect variegation in *Drosophila* on dose of a gene encoding an unusual zinc-finger protein. *Nature (Lond.)* **344**, 219–23.

Reznikoff, W.S., Siegele, D.A., Cowing, D.W. and Gross, C.A. (1985). The regulation of transcription initiation in bacteria. *Ann. Rev. Genet.* **19**, 355–87.

Rhodes, D. (1985). Structural analysis of a triple complex between the histone octamer, a *Xenopus* gene for 5S RNA and transcription factor IIIA. *EMBO J.* **4**, 3473–82.

Rhodes, D. and Klug, A. (1980). Helical periodicity of DNA determined by enzyme digestion. *Nature (Lond.)* **286**, 573–8.

Richard-Foy, H. and Hager, G.L. (1987). Sequence specific positioning of nucleosomes over the steroid-inducible MMTV promoter. *EMBO J.* **6**, 2321–8.

Richmond, T.J., Finch, J.T., Rushton, B., Rhodes, D. and Klug, A. (1984).

Structure of the nucleosome core particle at 7 Å resolution. *Nature (Lond.)* **311**, 532–7.

Riley, D. and Weintraub, H. (1979). Conservative segregation of parental histones during replication in the presence of cycloheximide. *Proc. Natl. Acad. Sci. USA* **76**, 328–32.

Ringertz, N.R. and Savage, R.E. (1976). *Cell Hybrids.* Academic Press, New York.

Ringertz, N.R., Nyman, U. and Bergman, M. (1985). DNA replication and H5 histone exchange during reactivation of chick erythrocyte nuclei in heter-okaryons. *Chromosoma* **91**, 391–6.

Robinson, G.W. and Hallick, L.M. (1982). Mapping the *in vivo* arrangement of nucleosomes on simian virus 40 with hydroxy methyltrimethylpsora-len. *J. Virol.* **41**, 78–87.

Rocha, E., Davie, J.R., van Holde, K.E. and Weintraub, H. (1984). Differential salt fractionation of active and inactive genomic domains in chicken erythrocyte. *J. Biol. Chem.* **259**, 8558–63.

Rodriquez-Campos, A., Shimamura, A. and Worcel, A. (1989). Assembly and properties of chromatin containing histone H1. *J. Molec. Biol.* **209**, 135–50.

Rose, S.M. and Garrard, W.T. (1984). Differentiation dependent chromatin alterations precede and accompany transcription of immunoglobulin light chain genes. *J. Biol. Chem.* **259**, 8534–44.

Roth, M.B. and Gall, J.G. (1987). Monoclonal antibodies that recognize tran-scription unit proteins on newt lampbrush chromosomes. *J. Cell Biol.* **105**, 1047–54.

Roth, S.Y. and Allis, C.D. (1992). Chromatin condensation: does histone H1 dephosphorylation play a role? *Trends Biochem. Sci.* **17**, 93–8.

Roth, S.Y., Schulman, I.G., Richman, R., Cook, R.G. and Allis, C.D. (1988). Characterization of phosphorylation sites in histone H1 in the amitotic macronucleus of *Tetrahymena* during different physiological states. *J. Cell Biol.* **107**, 2473–82.

Roth, S.Y., Dean, A. and Simpson, R.T. (1990). Yeast α2 repressor positions nucleosomes in TRP1/ARS1 chromatin. *Molec. Cell. Biol.* **10**, 2247–60.

Roth, S.Y., Shimizu, M., Johnson, L., Grunstein, M. and Simpson, R.T. (1992). Stable nucleosome positioning and complete repression by the yeast α2 repressor are disrupted by amino-terminal mutations in his-tones H4. *Genes Devel.* **6**, 411–425.

Rougvie, A.E. and Lis, J.T. (1988). The RNA polymerase II molecule at the 5′ end of the uninduced hsp 70 gene of D. melanogaster is transcriptionally engaged. *Cell* **54**, 795–804.

Rudolph, H. and Hinnen, A. (1987). The yeast PH05 promoter: Phosphate-control elements and sequences mediating mRNA start-site selection. *Proc. Natl. Acad. Sci. USA* **84**, 1340–44.

Ruiz-Carrillo, A., Jorcano, J.L., Eder, G. and Lurz, R. (1979). *In vitro* core particle and nucleosome assembly at physiological ionic strength. *Proc. Natl. Acad. Sci. USA* **76**, 3284–8.

Ryan, T.M., Rehringer, R.R., Martin, N.C., Townes, T.M., Palmiter, R.D. and Brinster, R.L. (1989). A single erythroid-specific DNaseI super-hypersensitive site activates high levels of human β-globin gene expression in transgenic mice. *Genes Devel.* **3**, 314–23.

Ryoji, M. and Worcel, A. (1984). Chromatin assembly in *Xenopus* oocytes: *in vivo* studies. *Cell* **37**, 21–32.

Saffer, L.D. and Miller, O.L. Jr. (1986). Electron microscopic study of *Saccharomyces cerevisiae* rDNA chromatin replication. *Molec. Cell. Biol.* **6**, 1147–57.

Sahasrabuddhe, C.G. and van Holde, K.E. (1974). The effect of trypsin on nuclease-resistant chromatin fragments. *J. Biol. Chem.* **249**, 152–6.

Satchwell, S.C., Drew, H.R. and Travers, A.A. (1986). Sequence periodicities in chicken nucleosome core DNA. *J. Molec. Biol.* **191**, 659–75.

Sawadogo, M. and Roeder, R.G. (1985). Interaction of a gene-specific transcription factor with the adenovirus major late promoter upstream of the TATA box region. *Cell* **43**, 165–75.

Schaffner, G., Schirm, S., Muller-Baden, B., Weber, F., and Schaffner, W. (1988). Redundancy of information in enhancers as a principle of mammalian transcriptional control. *J. Molec. Biol.* **201**, 81–90.

Schlissel, M.S. and Baltimore, D. (1989). Activation of immunoglobulin kappa gene rearrangement correlates with induction of kappa gene transcription. *Cell* **58**, 1001–7.

Schlissel, M.S. and Brown, D.D. (1984). The transcriptional regulation of *Xenopus* 5S RNA genes in chromatin: the roles of active stable transcription complexes and histone H1. *Cell* **37**, 903–11.

Schwabe, J.W.R., Neuhans, D. and Rhodes, D. (1990). Solution structure of the DNA-binding domain of the estrogen receptor. *Nature (Lond.)* **348**, 458–61.

Schwartz, D.C. and Cantor, C.R. (1984). Separation of yeast chromosome-sized DNAs by pulsed field gradient gel electrophoresis. *Cell* **37**, 67–75.

Sealy, L., Cohen, M. and Chalkley, R. (1986). *Xenopus* nucleoplasmin: egg vs oocyte. *Biochemistry* **25**, 3064–72.

Sedat, J. and Manuelidis, L. (1978). A direct approach to the structure of mitotic chromosomes. *Cold Spring Harbor Symp. Quant. Biol.* **42**, 331–50.

Segall, J., Matsui, T. and Roeder, R.G. (1980). Multiple factors are required for the accurate transcription of purified genes by RNA polymerase III. *J. Biol. Chem.* **255**, 11986–91.

Seidman, M.M., Levine, A.J. and Weintraub, H. (1979). The asymmetric segregation of paternal nucleosomes during chromosomal replication. *Cell* **18**, 439–49.

Senshu, T., Fukada, M. and Ohashi, M. (1978). Preferential association of newly synthesized H3 and H4 histones with newly synthesized replicated DNA. *J. Biochem. (Japan)* **84**, 985–8.

Serfling, E., Jasin, M. and Schaffner, W. (1985). Enhancers and eukaryotic gene transcription. *Trends Genet.* **1**, 224–30.

Sheehan, M.A., Mills, A.D., Sleeman, A.M., Laskey, R.A. and Blow, J.J. (1988). Steps in the assembly of replication-competent nuclei in a cell-free system from *Xenopus* eggs. *J. Cell Biol.* **106**, 1–12.

Shick, V.V., Belyavsky, A.V., Bavykin, S.G. and Mirzabekov, A.D. (1980).

Primary organization of the nucleosome core particles: sequential arrangement of histones along DNA. *J. Molec. Biol.* **139**, 491–517.

Shick, V.V., Belyavsky, A.V. and Mirzabekov, A.D. (1985). Primary organization of nucleosomes: interaction of non-histone high mobility group proteins 14 and 17 with nucleosomes, as revealed by DNA-protein cross-linking and immuno-affinity isolation. *J. Molec. Biol.* **185**, 329–59.

Shimamura, A. and Worcel, A. (1989). The assembly of regularly spaced nucleosomes in the *Xenopus* oocyte S150 extract is accompanied by de-acetylation of histone H4. *J. Biol. Chem.* **264**, 14524–30.

Shimamura, A., Tremethick, D. and Worcel, A. (1988). Characterization of the repressed 5S DNA minichromosomes assembled *in vitro* with a high-speed supernatant of Xenopus laevis oocytes. *Molec. Cell. Biol.* **8**, 4257–69.

Shimamura, A., Sapp, M., Rodriquez-Campos, A. and Worcel, A. (1989). Histone H1 represses transcription from minichromosomes assembled *in vitro*. *Mol. Cell Biol.* **9**, 5573–84.

Shimizu, M., Roth, S.Y., Szent-Gyorgi, C. and Simpson, R.T. (1991). Nucleosome are positioned with base pair precision adjacent to the α_2 operator in *Saccharomyces cerevisiae*. *EMBO J.* **10**, 3033–41.

Shrader, T.E. and Crothers, D.M. (1989). Artificial nucleosome positioning sequences. *Proc. Natl. Acad. Sci. USA* **86**, 7418–22.

Shrader, T.E. and Crothers, D.M. (1990). Effects of DNA sequence and histone-histone interactions on nucleosome placement. *J. Molec. Biol.* **216**, 69–84.

Sidik, K. and Smerdon, M.J. (1990). Nucleosome rearrangement in human cells following short patch repair of DNA damaged by bleomycin. *Biochemistry* **29**, 7501–11.

Simpson, R.T. (1978). Structure of the chromatosome, a chromatin core particle containing 160 base pairs of DNA and all the histones. *Biochemistry* **17**, 5524–31.

Simpson, R.T. (1990). Nucleosome positioning can affect the function of a cis-acting DNA element *in vivo*. *Nature (Lond.)* **343**, 387–9.

Simpson, R.T. (1991). Nucleosome positioning: occurrence, mechanisms and functional consequences. *Prog. Nucl. Acids Res. Molec. Biol.* **40**, 143–84.

Simpson, R.T. and Bergman, L.W. (1980). Structure of sea urchin sperm chromatin core particles. *J. Biol. Chem.* **255**, 10702–9.

Simpson, R.T. and Stafford, D.W. (1983). Structural features of a phased nucleosome core particle. *Proc. Natl. Acad. Sci. USA* **80**, 51–5.

Simpson, R.T., Thoma, F. and Brubaker, J.M. (1985). Chromatin reconstituted from tandemly repeated cloned DNA fragments and core histones: a model system for study of higher order structure. *Cell* **42**, 799–808.

Singer, D.S. and Singer, M.F. (1976). Studies on the interaction of histone H1 with superhelical DNA: characterization of the recognition and binding regions of H1 histone. *Nucl. Acids Res.* **3**, 2531–47.

Singh, J. and Rao, M.R.S. (1987). Interaction of rat testis protein, TP with nucleic acids *in vitro*. *J. Biol. Chem.* **262**, 734–40.

Smale, S.T., Schmidt, M.C., Berk, A.J. and Baltimore, D. (1990). Transcriptional activation by SP1 as directed through TATA or initiator: specific requirement for mammalian transcription factor IID. *Proc. Natl. Acad. Sci. USA* **87**, 4509–13.

Smerdon, M.J. and Thoma, F. (1990). Site-specific DNA repair at the nucleosome level in a yeast minichromosome. *Cell* **61**, 675–84.

Smerdon, M.J., Bedoyan, J. and Thoma, F. (1990). DNA repair in a small yeast plasmid folded into chromatin. *Nucl. Acids Res.* **18**, 2045–51.

Smith, P.A., Jackson, V. and Chalkley, R. (1984). Two-stage maturation process for newly replicated chromatin. *Biochemistry* **23**, 1576–81.

Smith, R.C., Dworkin-Rastl, E. and Dworkin, M.D. (1988). Expression of a histone H1–like protein is restricted to early *Xenopus* development. *Genes Devel.* **2**, 1284–95.

Smith, S. and Stillman, B. (1989). Purification and characterization of CAF-1 a human cell factor required for chromatin assembly during DNA replication *in vitro*. *Cell* **58**, 15–25.

Smith, S. and Stillman, B. (1991a). Stepwise assembly of chromatin during DNA replication *in vitro*. *EMBO J.* **10**, 971–80.

Smith, S. and Stillman, B. (1991b). Immunological characterization of chromatin assembly factor 1, a human cell factor required for chromatin assembly during DNA replication *in vitro*. *J. Biol. Chem.* **266**, 12041–7.

Sogo, J.M., Stahl, H., Koller, Th. and Knippers, R. (1986). Structure of replicating SV40 minichromosomes: The replication fork, core histone segregation and terminal structures. *J. Molec. Biol.* **189**, 189–204.

Solomon, M.J., Strauss, F. and Varshavsky, A. (1986). A mammalian high mobility group protein recognizes any stretch of six A–T base pairs in duplex DNA. *Proc. Natl. Acad. Sci. USA* **83**, 1276–80.

Solomon, M.J., Larsen, P.L. and Varsharsky, A. (1988). Mapping protein-DNA interactions *in vivo* with formaldehyde: evidence that histone H4 is retained on a highly transcribed gene. *Cell* **53**, 937–47.

Sopta, M., Burton, Z.F. and Greenblatt, J. (1989). Structure and associated DNA helicase activity of a general transcription intiation factor that binds to RNA polymerase II. *Nature (Lond.)* **341**, 410–14.

Srikantha, T., Landsman, D. and Bustin, M. (1988). Cloning of the chicken chromosomal protein HMG-14 cDNA reveals a unique protein with a conserved DNA binding domain. *J. Biol. Chem.* **263**, 13500–3.

Staynov, D.Z. and Crane-Robinson, C. (1988). Footprinting of linker histones H5 and H1 on the nucleosome. *EMBO J.* **7**, 3685–91.

Stedman, E. and Stedman, E. (1947). The chemical nature and functions of components of cell nuclei not histone but protein. *Cold Spring Harbor Symp.* **12**, 224–36.

Stein, A. and Bina, M. (1984). A model chromatin assembly system: factors affecting nucleosome spacing. *J. Molec. Biol.* **178**, 341–63.

Stein, A., Whitlock, J.P. and Bina, M. (1979). Acidic polypeptides can assemble both histones and chromatin *in vitro* at physiological ionic strength. *Proc. Natl. Acad. Sci. USA* **76**, 5000–4.

Sterner, R., Boffa, L., Chen, T.A. and Allfrey, V.G. (1987). Cell cycle-dependent changes in conformation and composition of nucleosomes containing human histone gene sequences. *Nucl. Acids Res.* **15**, 4375–91.

Stick, R. and Hansen, P. (1985). Changes in the nuclear lamina composition during early development of *Xenopus laevis*. *Cell* **41**, 191–200.

Stief, A., Winter, D.M., Stratting, W.E.H. and Sippel, A.E. (1989). A nuclear DNA attachment element mediates elevated and position independent gene activity. *Nature (Lond.)* **341**, 343–5.

Stillman, B. (1986). Chromatin assembly during SV40 DNA replication *in vitro*. *Cell* **45**, 555–65.

Stillman, B. (1989). Initiation of eukaryotic DNA replication *in vitro*. *Ann. Rev. Cell. Biol.* **5**, 197–245.

Straka, C. and Horz, W. (1991). A functional role for nucleosomes in the repression of a yeast promoter. *EMBO J.* **10**, 361–8.

Stringer, K.F., Ingles, C.J. and Greenblatt, J. (1990). Direct and selective binding of an acidic activation domain to the TATA-box factor TFIID. *Nature (Lond.)* **345**, 783–6.

Su, W., Jackson, S., Tjian, R. and Echols, H. (1991). DNA looping between sites for transcriptional activation: self association of DNA-bound Sp1. *Genes Devel.* **5**, 820–6.

Suau, P., Bradbury, E.M. and Baldwin, J.P. (1979). Higher-order structures of chromatin in solution. *Eur. J. Biochem.* **97**, 593–602.

Svaren, J. and Chalkley, R. (1990). The structure and assembly of active chromatin. *Trends Genet.* **6**, 52–6.

Szent-Gyorgi, C., Finkelstein, D.B. and Garrard, W.T. (1987). Sharp boundaries demarcate the chromatin structure of a yeast heat-shock gene. *J. Molec. Biol.* **193**, 71–80.

Tafuri, S.R. and Wolffe, A.P. (1990). *Xenopus* Y-box transcription factors: Molecular cloning, functional analysis and developmental regulation. *Proc. Natl. Acad. Sci. USA* **87**, 9028–32.

Takata, C., Albright, J.F. and Yomade, T. (1964). Lens antigens in a lens regenerating system studied by the immunofluorescent technique. *Devel. Biol.* **9**, 385–97.

Talbot, D. and Grosveld, F. (1991). The 5′ HS 2 of the globin locus control region enhances transcription through the interaction of a multimeric complex binding at two functionally distinct NF-E2 binding sites. *EMBO J.* **10**, 1391–8.

Talbot, D., Philipsen, S., Fraser, P. and Grosveld, F. (1990). Detailed analysis of the site 3 region of the human β-globin dominant control region. *EMBO J.* **9**, 2169–78.

Taylor, I.C.A., Workman, J.L., Schmetz, T-J. and Kingston, R.E. (1991). Facilitated binding of GAL4 and heat shock factor to nucleosomal templates: differential function of DNA-binding domains. *Genes Devel.* **5**, 1285–98.

Theulaz, I., Hipskind, R., TenHeggeler-Bordier., B., Green, S., Kumar, V., Chambon, P., and Wahli, W. (1988). Expression of human estrogen receptor mutants in *Xenopus* oocytes: correlation between transcriptional activity and ability to form protein-DNA complexes. *EMBO J.* **7**, 1653–60.

Thoma, F. (1986). Protein-DNA interactions and nuclease sensitive regions determine nucleosome positions on yeast plasmid chromatin. *J. Molec. Biol.* **190**, 177–90.

Thoma, F. (1991). Structural changes in nucleosomes during transcription: strip, split or flip? *Trends Genet.* **7**, 175–77.

Thoma, F. and Simpson, R.T. (1985). Local protein-DNA interactions may determine nucleosome positions on yeast plasmids. *Nature (Lond.)* **315**, 250–3.

Thoma, F. and Zatchej, M. (1988). Chromatin folding modulates nucleosome positioning in yeast minichromosomes. *Cell* **55**, 945–53.

Thoma, F., Koller, T. and Klug, A. (1979). Involvement of histone H1 in the organization of the nucleosome and the salt-dependent superstructures of chromatin. *J. Cell Biol.* **83**, 402–27.

Thoma, F., Bergman, L. W. and Simpson, R.T. (1984). Nuclease digestion of circular TRP1ARS1 chromatin reveals positioned nucleosomes separated by nuclease sensitive regions. *J. Molec. Biol.* **177**, 715–33.

Thomas, G.P. and Mathews, M.B. (1980). DNA replication and the early to late transition in adenovirus infection. *Cell* **22**, 523–33.

Travers, A.A. (1989). DNA conformation and protein binding. *Ann. Rev. Biochem.* **58**, 427–52.

Travers, A.A. and Klug, A. (1987). The bending of DNA in nucleosomes and its wider implications. *Phil. Trans. R. Soc. Lond. B* **317**, 537–61.

Tremethick, D., Zucker, D. and Worcel, A. (1990). The transcription complex of the 5S RNA gene, but not the transcriptional factor TFIIIA alone, prevents nucleosomal repression of transcription. *J. Biol. Chem.* **265**, 5014–23.

Tsanev, R. and Sendov, B. (1971). Possible molecular mechanism for cell differentiation in multicellular organisms. *J. Theor. Biol.* **30**, 337–93.

Tuan, D., Solomon, W., Li, Q. and London, I.M. (1985). The 'β-like-globin' gene domain in human erythroid cells. *Proc. Natl. Acad. Sci. USA* **82**, 6384–8.

Tullius, T.D. and Dombroski, B.A. (1985). Iron (II) EDTA used to measure the helical twist along any DNA molecule. *Science* **230**, 679–81.

Turner, B.M. (1991). Histone acetylation and control of gene expression. *J. Cell Sci.* **99**, 13–20.

Ulitzer, N. and Gruenbaum, Y. (1989). Nuclear envelope assembly around sperm chromatin in cell-free preparations from *Drosophila* embryos. *FEBS Lett.* **259**, 113–16.

Van Dyke, M.W., Roeder, R.G. and Sawadogo, M. (1988). Physical analysis of transcription preinitiation complex assembly on a class II gene promoter. *Science* **241**, 1335–8.

van Holde, K.E. (1989). *Chromatin*. Springer-Verlag, New York.

Varshavsky, A.J., Bakayev, V.V. and Georgiev, G.P. (1976). Heterogeneity of chromatin subunits *in vitro* and location of histone H1. *Nucl. Acids Res.* **3**, 477–92.

Varshavsky, A.J., Sundin, O.H. and Bohn, M.J. (1978). SV40 viral minichromosome: preferential exposure of the origin of replication as probed by restriction endonucleases. *Nucl. Acids. Res.* **5**, 3469–77.

Vernet, G., Sala-Rovira, M., Maeder, M., Jacques, F. and Herzog, M. (1990). Basic nuclear proteins of the histone less eukaryote *Gypthecodinium cohnii*

(Pyrrhophyta): two dimensional electrophoresis and DNA binding properties. *Biochim. Biophys. Acta* **1048**, 281–9.

Voeller, B.R. (1968). *The Chromosome Theory of Inheritance: Classic Papers in Development and Hereditary.* Appleton, New York.

Wabl, M.R. and Burrows, P.D. (1984). Expression of immunoglobulin heavy chain at a high level in the absence of a proposed immunoglobulin enhancer in cis. *Proc. Natl. Acad. Sci. USA* **81**, 2452–5.

Wakefield, L. and Gurdon, J.B. (1983). Cytoplamic regulation of 5S RNA genes in nuclear-transplant embryos. *EMBO J.* **2**, 1613–19.

Walker, J., Chen, T.A., Sterner, R., Berger, M., Winston, F. and Allfrey, V.G. (1990). Affinity chromatography of mammalian and yeast nucleosomes: two modes of binding of transcriptionally active mammation nucleosomes to organomercurial columns and contrasting behaviour of the active nucleosomes of yeast. *J. Biol. Chem.* **265**, 5736–46.

Wang, X.F. and Calame, K. (1986). SV40 enhancer-binding factors are required at the establishment but not the maintenance step of enhancer-dependent transcriptional activation. *Cell* **47**, 241–7.

Wasylyk, B. and Chambon, P. (1979). Transcription by eukaryotic RNA polymerases A and B of chromatin assembled *in vitro*. *Eur. J. Biochem.* **98**, 317–27.

Watson, J.D. and Crick, F.H.C. (1953). A structure for deoxyribosenucleic acids. *Nature (Lond.)* **171**, 737–8.

Weintraub, H. (1984). Histone H1-dependent chromatin superstructures and the suppression of gene activity. *Cell* **38**, 17–27.

Weintraub, H. (1985). Assembly and propagation of repressed and de-repressed chromosomal states. *Cell* **42**, 705–11.

Weintraub, H. (1988). Formation of stable transcription complexes as assayed by analysis of individual templates. *Proc. Natl. Acad. Sci. USA* **85**, 5819–23.

Weintraub, H. and Groudine, M. (1976). Chromosomal subunits in active genes have an altered conformation. *Science* **193**, 848–56.

Weintraub, H., Worcel, A. and Alberts, B. (1976). A model for chromatin based upon two symmetrically paired half nucleosomes. *Cell* **9**, 409–17.

Weintraub, H., Beug, H., Groudine, M. and Graf, T (1982). Temperature sensitive changes in the structure of globin chromatin in lines of red cell precursors transformed by ts-AEV. *Cell* **28**, 931–40.

Weisbrod, S., Wickens, M.P., Whytock, S. and Gurdon, J.B. (1982). Active chromatin of oocytes injected with somatic cell nuclei or cloned DNA. *Devel. Biol.* **94**, 216–29.

Weiss, E., Ghose, D., Schultz, P. and Oudet, P. (1985). T-antigen is the only detectable protein on the nucleosome-free origin region of isolated simian virus 40 minichromosomes. *Chromosoma (Berl.)* **92**, 391–400.

White, J.H., Cozzarelli, N.R. and Bauer, W.R. (1988). Helical repeat and linking number of surface wrapped DNA. *Science* **241**, 323–7.

White, M.J.D. (1973). *Animal Cytology and Evolution.* Cambridge University Press, Cambridge, pp. 1–58.

Widom, J. and Klug, A. (1985). Structure of the 300 Å chromatin filament: X-ray diffraction from orientated samples. *Cell* **43**, 207–13.

Wildeman, A.G., Zenke, M., Schatz, C., Wintzerith, M., Grundstrom, T., T. Matthes, H., Takahaski, K. and Chambon, P. (1986). Specific protein binding to the simian virus 40 enhancer *in vitro. Molec. Cell. Biol.* **6**, 2098–105.

Williams, S.P. and Langmore, J.P. (1991). Small angle x-ray scattering of chromatin. *Biophys. J.* **69**, 606–18.

Williamson, P. and Felsenfeld, G. (1978). Transcription of histone-covered T7 DNA by *Escherichia coli* RNA polymerase. *Biochemistry* **17**, 5695–705.

Williamson, R. (1970). Properties of rapidly labelled deoxyribonucleic acid fragments isolated from the cytoplasm of primary cultures of embryonic mouse liver cells. *J. Molec. Biol.* **51**, 157–68.

Wilson, E.B. (1925). *The Cell in Development and Heredity.* Macmillan, New York.

Wingender, E., Jahn, D. and Seifart, K.H. (1986). Association of RNA polymerase III with transcription factors in the absence of DNA. *J. Biol. Chem.* **261**, 1409–13.

Wolffe, A.P. (1988). Transcription fraction TFIIIC can regulate differential *Xenopus* 5S RNA gene transcription *in vitro. EMBO J.* **7**, 1071–9.

Wolffe, A.P. (1989a). Transcriptional activation of *Xenopus* class III genes in chromatin isolated from sperm and somatic nuclei. *Nucl. Acids Res.* **17**, 767–80.

Wolffe, A.P. (1989b). Dominant and specific repression of Xenopus oocyte 5S RNA genes and satellite I DNA by histone H1. *EMBO J.* **8**, 527–37.

Wolffe, A.P. (1990a). New approaches to chromatin function. *New Biol.* **2**, 211–18.

Wolffe, A.P. (1990b). Transcription complexes. *Prog. Clin. Biol. Res.* **322**, 171–86.

Wolffe, A.P. (1991a). Developmental regulation of chromatin structure and function. *Trends Cell Biol.* **1**, 61–6.

Wolffe, A.P. (1991b). RNA polymerase III transcription. *Curr. Opin. Cell Biol.* **3**, 461–6.

Wolffe, A.P. and Brown, D.D. (1986). DNA replication *in vitro* erases a Xenopus 5S RNA gene transcription complex. *Cell* **47**, 217–27.

Wolffe, A.P. and Brown, D.D. (1987). Differential 5S RNA gene expression *in vitro. Cell* **51**, 733–40.

Wolffe, A.P. and Brown, D.D. (1988). Developmental regulation of two 5S ribosomal RNA genes. *Science* **241**, 1626–32.

Wolffe, A.P. and Drew, H.R. (1989). Initiation of transcription on nucleosomal templates. *Proc. Natl. Acad. Sci. USA* **86**, 9817–21.

Wolffe, A.P. and Morse, R.H. (1990). The transcription complex of the *Xenopus* somatic 5S RNA gene. *J. Biol. Chem.* **265**, 4592–9.

Wolffe, A.P. and Schild, C. (1991). Chromatin assembly. *Methods Cell Biol.* **36**, 541–59.

Wolffe, A.P., Jordan, E. and Brown, D.D. (1986). A bacteriophage RNA polymerase transcribes through a *Xenopus* 5S RNA gene transcription complex without disrupting it. *Cell* **44**, 381–9.

Wood, W.I. and Felsenfeld, G. (1982). Chromatin structure of the chicken β-globin gene region: sensitivity to DNaseI, micrococcal nuclease, and DNaseII. *J. Biol. Chem.* **257**, 7730–6.

Woodcock, C.L., Frado, L.L. and Rattner, J.B. (1984). The higher-order structure of chromatin: evidence for a helical ribbon arrangement. *J. Cell Biol.* **99**, 42–52.

Woodland, H.R. and Adamson, E.D. (1977). The synthesis and storage of histones during the oogenesis of *Xenopus laevis*. *Devel. Biol.* **57**, 118–35.

Woodland, H.R., Flynn, J.M. and Wyllie, A.J. (1979). Utilization of stored mRNA in Xenopus embryos and its replacement by newly synthesized transcripts: histone H1 synthesis using interspecies hybrids. *Cell* **18**, 165–71.

Worcel, A. (1978). Molecular architecture of the chromatin fiber. *Cold Spring Harbor Symp. Quant. Biol.* **42**, 313–24.

Worcel, A. and Burgi, E. (1972). On the structure of the folded chromosome of *E. coli*. *J. Molec. Biol.* **71**, 127–48.

Worcel, A., Han, S. and Wong, M.L. (1978). Assembly of newly replicated chromatin. *Cell* **15**, 969–77.

Workman, J.L. and Roeder, R.G. (1987). Binding of transcription factor TFIID to the major late promoter during *in vitro* nucleosome assembly potentiates subsequent initiation by RNA polymerase II. *Cell* **51**, 613–22.

Workman, J.L., Abmayr, S.M., Cromlish, W.A. and Roeder, R.G. (1988). Transcriptional regulation of the immediate early protein of pseudorabies virus during *in vitro* nucleosome assembly. *Cell* **55**, 211–19.

Workman, J.L., Roeder, R.G. and Kingston, R.E. (1990). An upstream transcription factor, USF (MLTF), facilitates the formation of preinitiation complexes. *EMBO J.* **9**, 1299–308.

Workman, J.L., Taylor, I.C.A. and Kingston, R.E. (1991). Activation domains of stably bound GAL4 derivaties alleviate repression of promoters by nucleosomes. *Cell* **64**, 533–44.

Wormington, W.M. and Brown, D.D. (1983). Onset of 5S RNA gene regulation during *Xenopus* embryogenesis. *Devel. Biol.* **99**, 248–57.

Wormington, W.M., Schlissel, M. and Brown, D.D. (1982). Developmental regulation of *Xenopus* 5S RNA genes. *Cold Spring Harbor Symp Quant. Biol.* **47**, 879–84.

Wu, C. and Gilbert, W. (1981). Tissue-specific exposure of chromatin structure at the 5′ terminus of the rat prepro insulin II gene. *Proc. Natl. Acad. Sci. USA* **78**, 1577–80.

Wu, C., Binham, P.M., Livak, K.J., Holmgren, R. and Elgin, S.C.R. (1979). The chromatin structure of specific genes: evidence for higher order domains of defined DNA sequence. *Cell* **16**, 797–806.

Wyllie, A.H., Gurdon, J.B. and Price, J. (1977). Nuclear localization of an oocyte component required for the stability of injected DNA. *Nature (Lond.)* **268**, 150–2.

Wyllie, A.H., Laskey, R.A., Finch, J. and Gurdon, J.B. (1978). Selective DNA conservation and chromatin assembly after injection of SV40 DNA into *Xenopus* oocytes. *Devel. Biol.* **64**, 178–88.

Xu, M., Barnard, M.B., Rose, S.M., Cockerill, P.N., Huang, S-Y. and Garrard, W.T. (1986). Transcription termination and chromatin structure of the active immunoglobulin K gene locus. *J. Biol. Chem.* **261**, 3838–45.

Yamamoto, K.R. (1985). Steroid receptor regulated transcription of specific genes and gene networks. *Ann. Rev. Genet.* **19**, 209–52.

Yanagi, K., Privé, G.G. and Dickerson, R.E. (1991). Analysis of local helix geometry in three B-DNA decamers and eight dodecamers. *J. Molec. Biol.* **217**, 201–14.

Yang, C.C. and Nash, H.A. (1989). The interaction of *E. coli* IHF protein with its specific binding sites. *Cell* **57**, 869–80.

Yao, J., Lowary, P.T. and Widom, J. (1990). Direct detection of linker DNA bending in defined-length oligomers of chromatin. *Proc. Natl. Acad. Sci. USA* **87**, 7603–7.

Yao, J., Lowary, P.T. and Widom, J. (1991). Linker DNA bending by the core histones of chromatin. *Biochemistry* **30**, 8408–14.

Zakian, V.A. (1989). Structure and function of telomeres. *Ann. Rev. Genet.* **23**, 579–604.

Zaret, K.S. and Yamamoto, K.R. (1984). Reversible and persistent changes in chromatin structure accompany activation of a glucocorticoid dependent enhancer element. *Cell* **38**, 29–38.

Zenke, M., Grundstrom, T., Matthes, H., Wintzerith, M., Schatz, C., Wildeman, A. and Chambon, P. (1986). Multiple sequence motifs are involved in SV40 enhancer function. *EMBO J.* **5**, 387–97.

Zentgraf, H. and Franke, W.W. (1984). Differences of supra nucleosomal organization in different kinds of chromatin: cell type-specific globular subunits containing different numbers of nucleosomes. *J. Cell Biol.* **99**, 272–86.

Zink, B. and Paro, R. (1989). *In vivo* binding pattern of a *trans* regulator of homeotic genes in *Drosophila melanogaster*. *Nature (Lond.)* **337**, 468–71.

Zweidler, A. (1980). Nonallelic histone variants in development and differentiation. *Devel. Biochem.* **15**, 47–56.

Index

A

A elements, 146
Abd-A/abd-B gene, 146–7
Acanthamoeba spp., transcription, 110, 113
Acetylation of histones, 54–6, 86–7, 132–3, 161, 162
Adenine, *see also* Oligo(dA.dT) tracts
  base pairing, 5, 7
  structure, 5
Adenosine triphosphate, chromatin assembly requiring, 94
A-DNA, 8
ADP-ribosylation (poly(ADP-ribose) polymerase activity), 160
  core histone, 57
α-mating types, 84, 148–9
α₂-repressor, 84
α-satellite family of DNA sequences, 47
  heterochromatin associated with, 65
Arrays, nucleosomal, compaction/folding, 31–3, 34–6, 168
ARS1 replication origin, 82, 83, 139
Assembly, 68–99
  chromatin, 73–96, 168
    in vitro, 88–96
    in vivo, 73–88
  nucleus, 96–9

ATP, chromatin assembly requiring, 94
Att P site, 102

B

Bacteriophage λ DNA
  replication, 101–2
  in *Xenopus* eggs, assembly into chromatin
    in vitro, 97–8
    in vivo, 76–7
Balbani ring genes, 50–1
Bases
  accessibility in nucleosome, studies of, 23–4, 26
  pairing, 5, 7
  stacking, 5, 7
B-DNA, 6, 10
Beads-on-a-string configuration of chromatin, 34, *see also* Superbeads
Bithorax complex genes, 146–7
Bovine papilloma virus, episomes based on, 80

C

Caenorhabditis elegans, DNA replication and transcription during development, 159

CAF-I, 96
Cdc 2 mitotic kinase, 52, 60
CENP-A, 46–7, 47
CENP-B, 46–7, 47, 67
CENP-C, 46–7, 47
Centromere
  heterochromatin at, 64–5, 65
  proteins associated with DNA at,
    46–7, 64–5
Chironomus tentans, Balbani ring
    genes, 50–1
Chromatin assembly factor-I, 96
Chromomeres, 49
Chromosomes, structure/
    architecture, 39–67, 140–7
    modulation, 51–67
Circular DNA, closed, *see* Closed
    circular DNA
Class I/II/III genes, *see* Gene(s)
Closed circular DNA
  chromatin structures on,
    detection, 74, 75
  structure, 8–9
Coiled linker model of chromatin
    fiber, 37, 38
Cold-shock domain of Y-box, 121
Compaction, *see* Arrays,
    nucleosomal; Folding
Condensation of DNA/chromatin/
    chromosome, *see* Folding
Core histones, 11–13, 51–2, 54–7
  in nucleosome
    octamer of, 16
    position of, 19–20, 23–8, 28–31
  post-translational modification,
    53, 54–7, 132–3, 161, 162
  variants, 12–13, 51–2, 53, 62
Core particles, nucleosomal, 16, 18–
    20
  hydroxyl radical cleavage, 21–2
  organization, 18–20
Cytoplasm and nuclear structure,
    interactions, 68–73
Cytosine
  base pairing, 5, 7
  structure, 5

D
Deacetylase inhibitors, 86–7
  histone hyperacetylation induced
    with, 55

Deacetylation of histones, 86–7
Development
  RNA polymerase accessibility to
    chromatin through, changes
    in, 125
  transcription varying with, 125,
    158–9
  histone, 51–4
Dimethylsulfate, studies of base
    accessibility in nucleosome
    employing, 23–4, 26
Disc-like shape dimensions of
    nucleosomes, 20
DNA, 5–11
  cleavage reagents, studies
    employing, 8, 21, 23, *see also*
    specific cleavage reagents
  closed circular, *see* Closed circular
    DNA
  exogenous, chromatin assembly
    on, 73–88
  general structure, 5–11
  linker, *see* Linker DNA
  nucleosomal
    organization, 16–20
    positioning of nucleosome and
      the role of, 28–31
    structure, 20–3
  proteins binding to, *see* Non-
    histone/DNA-binding
    proteins
  repair, 159–60
    ADP-ribosylation and, 57, 160
    chromatin structure and, 57,
      159–60
  replication/synthesis, *see*
    Replication
  transcription, *see* Transcription
DnaA protein, 101
DnaC protein, 101
DNaseI, 140–7
  hypersensitivity to, 141–7
  sensitivity to, 140–1
    DNA structural studies
      employing, 8, 18, 19
    of nucleosome core particle, 18,
      19
Drosophila spp.
  DNA replication and
    transcription during
    development, 158

DNaseI-hypersensitive sites in
chromatin of, 142, 143, 146–7
histone genes, organizations into
chromatin, 74–6
Drug resistance markers,
chromosome integration
studies employing, 81

E
Eggs, *Xenopus*
chromatin assembly in, 73–8
nuclear assembly in extracts of,
96–9
nucleosome assembly in extracts
of, 90–5
somatic cell nuclei transplanted
into, 69–71
Electric dichroism, nucleosome
array studies employing, 36
Electrophoresis, gel, *see* Gel
electrophoresis
Enhancers, 105–6, 114, 118, 137,
146
Epstein-Barr virus, 104
Erythrocyte
histone phosphorylation and
mitosis in developing,
uncoupling, 59–60
nucleus, heterochromatinized, 65
Escherichia coli
DNA replication, 101, 104–5
RNA polymerase, *see* RNA
polymerase, *E. coli*

F
Fiber, chromatin, 31–8
histone exchange into/out of, 72,
137–8
nucleosome organization into,
31–8
Folding of DNA/chromatin/
chromosome, 31–2, 34–6,
168
histone post-translational
modification and,
uncoupling, 58–60
into nucleosomes, chromosome
structure and, 39–43
topoisomerase role, 45–6

G
Gel electrophoresis
non-denaturing, nucleosome
structural studies employing,
17–18
pulsed-field, chromatin/
chromosome structural
studies employing, 40–1
Gene(s), *see also specific genes*
class I, *see* ribosomal RNA; RNA
polymerase I
class II, *see* RNA polymerase II
class III, *see* RNA (5S and
transfer); RNA polymerase II
transcription/activity, *see*
Transcription
β-Globin genes, DNaseI
hypersensitive sites, 142–3,
144–5
Globular clusters of nucleosomes, 36
Globular histone domain/region, 13
DNA organization in
nucleosomes and the, 25, 27
Glucocorticoid receptor, gene
expression activated by, 80,
118, 135–6, 137
Guanine
base pairing, 5, 7
structure, 5

H
H1, 13–14, 162–3
class III gene expression and, 125
deficiency in transcriptionally
active chromatin, 162–3, 164
deposition on nascent DNA, 86
exchange into/out of chromatin
fiber, 72, 137–8
nucleosomal structure/integrity/
assembly and, 18, 31–3, 38,
129, 131
post-translational modification,
57–9, 60, 87
variants, 52–4
H2A, 11
amino acid sequence, 12
deposition/sequestration on
nascent DNA, 152, 153, 154
exchange into/out of chromatin
fiber, 137–8, 162

H2A (*cont.*)
  H1 and, linkage/contact between,
    33
  in nucleosome, position, 25, 88,
    94, 95, 96
  post-translational modification,
    57
  variants, 52, 53, 87
H2A/H2B, 25, 27–8
  dissociation/reassociation, 27–8
  nucleoplasmin and, association,
    91
  role, 28
H2B, 11, *see also* H2A/H2B
  amino acid sequence, 12
  deposition/sequestration on
    nascent DNA, 86, 152, 153,
    154
  exchange into/out of chromatin
    fiber, 137–8, 162
  in nucleosome, position, 25, 82,
    94, 95, 96
  post-translational modification,
    57
  variants, 52, 53, 87
H3, 11
  amino acid sequence, 11–12, 12
  CENP-A and, homology
    between, 47
  deposition/sequestration onto
    nascent DNA, 86, 152, 153,
    154
  in nucleosome, position, 23–5,
    88, 96
  post-translational modification,
    55, 56–7
  sulfhydryl residues, accessibility,
    162
    in nuclease-sensitive
      chromatin, 141
H3/H4 tetramer, 23–5
  dissociation/reassociation, 28
  N1/N2 and, association, 91
  role, 23, 28
H4, 11, *see also* H3/H4 tetramer
  amino acid sequence, 11–12, 12
  deacetylation, 86, 87
  deletion in tail of, effects, 149
  in nucleosome, position, 8, 23–5,
    93
  sequestration onto nascent DNA,

    152, 153, 154
H5, 13–14
  erythrocyte nuclear transcription
    suppressed by, 65
  post-translational modification,
    59–60
Heat shock proteins, DNaseI-
    sensitive/hypersensitive
    sites, 143, 146, 162
Heat shock transcription factor,
    130, 143–4
Helix, DNA, *see also* Superhelices
  dimensions, 6, 8, 10
  folding into nucleosomes,
    chromosome structure and,
    39–43
  nucleosomal, changes in
    periodicity/deformation, 20,
    20–3
Helix-turn-helix proteins, 120
Herpes simplex virus thymidine
    kinase promoter, 105–6
Heterochromatin, 63–5
Heterochromatin protein-1, 64
Heterokaryons, 71–3
High mobility group (HMG)
    protein(s), 66
  type 1 (HMG1/I), 47, 66, 66–7, 67
  type 2 (HMG2), 66, 66–7, 67
  type 14 (HMG14), 66, 67
  type 17 (HMG17), 66, 67
Histones, 3, 11–14, 51–61, 168–9,
    see also individual histones
  core, *see* Core histones
  DNA replication associated with
    synthesis/sequestration of,
    81–7, 152–3, 154
  exchange, into/out of chromatin,
    72, 137–8, 162
  genes, organization into
    chromatin, 74–6
  linker, *see* Linker histones
  octamers, 16, 89, 138
  post-translational modification,
    53, 54–61, 86–7, 132–3, 161,
    162
  structure, 11–14
  variants, 12–13, 51–4, 62, 87
HMG proteins, *see* High mobility
    group proteins
HP1, 64

Hsp26 gene, DNaseI-hypersensitive sites, 143
Hsp70 gene, DNaseI-hypersensitive sites, 143
Hsp82 gene, DNaseI-sensitive sites, 162
HSTF, 130, 143–4
HU protein, 101, 102
Hydrogen bonding, 5, 7
Hydroxyl radicals, DNA structural studies employing, 8, 19, 21, 23
Hyperacetylation, core histone, 55

I
IHF, 102
Immediate-early genes, transcription, 141
Immunoglobulin H (heavy chain) enhancer, 114
Immunoglobulin K gene, 81
INCENPs, 46
Int, 102
Intasome, 102
Integration host factor, 102

K
Kinase, mitotic, sea urchin, 52, 60

L
Lac repressor, 132
λ DNA, *see* Bacteriophage λ DNA
Lamin A, 43, 44
Lamin B, 43
Lamin C, 43, 44
Lamin L$_{111}$, 98
Lamina, nuclear, 43–4
  assembly, 98
Lampbrush chromosomes, 48–9
LCRs, 82, 145
Leucine zipper, 121
Linker DNA
  coiling between adjacent nucleosomes, 37, 38
  H1 binding to, 32
  non-histone proteins binding to, 124
Linker histones, 12–14, 52–4

erythrocyte nuclear transcription suppressed by, 65
phosphorylation, 57–61, 87
variants, 52–4, 62
Linking number paradox, 22–3
Locus control regions, 82, 145
Loop, radial, as model of chromosome structure, 39–43
Looping-out hypothesis, 105

M
Magnesium, chromatin assembly requiring, 94
Mammalian cell extracts, nucleosome assembly in, 95–6
MARs, 45, 145–6
Mating type, yeast, 84, 148–9
Matrix (scaffold), nuclear, 43–8, 60–1, 168
Matrix attachment regions, 45, 145–6
MCM1 protein, 84, 148–9
Membrane, nuclear, assembly, 98
Methylation, core histone, 57
Micrococcal nuclease, nucleosomal studies employing, 16–18, 37
Minichromosomes, 74
  SV40, 78–80, 142
  yeast, 82–5, 139–40, 148
Mitosis
  chromosome compaction and, uncoupling, 58–60
  linker histone phosphorylation and, uncoupling, 58–60
Mitotic kinase, 52, 60
Mouse mammary tumor virus, 135–7
  enhancer, 118
  glucocorticoid receptor binding and, 118, 135–6, 137
  long terminal repeat, 135
MPF kinase, 52, 60

N
N1/N2, 67, 91–2
NF1/CTF, 48, 136
Non-histone/DNA-binding proteins, 3, 63–7, 111, 120–2,

N-h/DNA-b proteins (*cont.*)
  see also Trans-acting factors
  non-specific, 123–6
  sequence-specific, 120–2, 126–40
N-terminal tails of core histones
  deletion, effects, 149
  post-translational modification,
    54–5, 56–7
Nuclease, *see specific nuclease*
Nucleoplasmin, 90–2, 94
  phosphorylation, 94
Nucleosome, 14–38, 90–5, 155–9
  assembly in *Xenopus* egg/oocyte
    extracts, 90–5
  structure/organization, 14–38, 75,
    129–30, 148, 155–9, 167–8
  into chromatin fiber, 31–8
  DNA replication affecting,
    155–9
  DNA transcription and, 163,
    164–6
Nucleosome hypothesis/model,
  14–15
Nucleosome structures, specific
  term known as, 28
Nucleotide, structure, 6
Nucleus, 68–166
  assembly *in vitro*, 96–9
  cytoplasm and structure of,
    interactions, 68–73
  heterochromatinized, 65
  processes, 100–66
    occurring in chromatin, 100–66
    overview, 100–23
    problems for, 122–3
  scaffold/matrix, 43–8, 60–1, 168
  transplantation experiments,
    69–71

O
O protein, 101, 103
Oligo(dA.dT) tracts, structure, 10
Oncogenes, proto-, transcription,
  141
Oocytes, *Xenopus*
  chromatin assembly in, 71–8
  nucleosome assembly in extracts
    of, 90–5
  S150 system, 93, 94
  somatic cell nuclei transplanted

  into, 69–71
ORI region, SV40, 78–9, 142
OriC, 101
O-some, 102, 103

P
Papilloma virus, bovine, episomes
  based on, 80
Phage λ DNA, *see* Bacteriophage λ
  DNA
PHO2 protein, 134
PHO4 protein, 134
PHO5 gene, 133–5
N-(5′-Phosphoribosyl)anthranilate
  isomerase gene (TRP1), 82,
  83
Phosphorylation
  core histone, 56–7
  linker histone, 57–61, 87
  nucleoplasmin, 94
Plasmid DNA in *Xenopus* eggs
  cytoplasm, assembly into
    chromatin, 76
Poly(ADP-ribose) polymerase
  activity, *see* ADP-ribosylation
Polynucleotide chain, structure/
  arrangement, 6, 7
Polytene chromosomes, 49–60
Post-translational modification of
  histones, 53, 54–61, 86–7,
  132–3, 161, 162
POU transcription factors, 67
Prokaryote
  DNA
    replication, 101–2, 104–5
    in *Xenopus* eggs, assembly into
      chromatin, *in vitro*, 76–7,
      97–8
  RNA polymerase, 164, 165
    nucleosomal DNA accessibility
      to, 124, 164
Promoters, 105–6, 118
Protamines, chromatin remodelling
  and, 61
Proteins, *see also specific proteins*
  in chromosome scaffold
    preparations, 43–8
  non-histone/DNA-binding, *see*
    Non-histone proteins
Proto-oncogenes, transcription, 141

Pulsed-field gel electrophoresis, chromatin/chromosome structural studies employing, 40–1

R
Radial loop model of chromosome structure, 39–43
RAP 1, 48
Repair, DNA, *see* DNA
Replication, DNA, 85–8, 101–5, 151–9
    chromatin assembly during, 85–8
    initiation, 157–8
    transcription and the effects of, 118, 119, 151–9
Replication fork, 156–7
Research, chromatin, development, 2–3
Ribbon-like structure of chromatin, 34, 36
Ribosomal RNA
    5S, gene for
        nucleosome position and the, 29–30
        organization into chromatin of, 74–84
    transcription, 110, 113, 160
RNA
    5S, genes, assembly, 154
    ribosomal, *see* Ribosomal RNA
    transcription, 107, 108, 109, 113, 115, 129, 132–3, 154–5, 157
    histone-coding, modification of translation product, 53, 54–61, 86–7
    synthesis, *see* transcription
    transfer, transcription, 107, 109
RNA polymerase, *E. coli*, 164
    nucleosomal DNA accessibility to, 124, 164
RNA polymerase, SP6, 165
RNA polymerase, T7, 165
RNA polymerase I, 107, 160, 165
RNA polymerase II, 106–7, 160–1, 163–4, 165, 165–6
RNA polymerase III, 107, 108, 125, 165
Rod-like structure of chromatin, 34, 35
Rotational position of nucleosomes, 29

S
S150 system, *Xenopus* oocyte, 93, 94
Saccharomyces spp., *see* Yeast
Salt, nucleosome association and effects of, 88–90
SARs, 45, 145–6
Scaffold, nuclear, 43–8, 60–1, 168
Scaffold attachment regions, 45, 145–6
Scaffold proteins, 44–5, 60–1
Scaffolding hypothesis, 42
Sc I/II/III, 44, 61
SCS elements, 146
Sea urchin
    DNA replication and transcription during development, 159
    histone variants, 51–2
    mitotic kinase, 52
    spermatogenesis, histone phosphorylation and chromatin compaction in, 60
SL1, 110
Solenoidal structure, nucleosome chains wound into, 35, 36, 38
Somatic cells
    chromatin assembly on DNA introduced into, 78–82
    fusion (= heterokaryons), 71–3
    histone synthesis coupled to DNA replication in, 85–7
    nuclei of, *Xenopus* eggs oocytes microinjected with, 69–71
SP1, 104, 105–6, 107
SP6 RNA polymerase, 165
Specialized chromatin structure elements, 146
Spermatogenesis, chromatin structural changes during, 60, 61–3
S-phase, DNA replication during, 157–8
    histone synthesis coupled to, 85
SPKK (S/T P-X-K/R) motifs in histone tail regions, 52, 58
Structure
    of chromatin, 1–2, 1–67, 140–66, 167–8
    DNA repair and, 159–60
    dynamic nature, 72

Structure (*cont.*)
    local organization, *trans*-acting
      factors and, 147–50
    processive enzyme complexes
      and, 150–66
    nuclear, cytoplasm and,
      interactions, 68–73
Sulfhydryl residues, H3, *see* H3
Superbeads, 36
Superhelices
    in closed circular DNA, 8–9
    nucleosomal, 23
    removal by topoisomerase, 41
Supvar(3)7, 64
SV40, 73–4, 78–80, 95–6, 103–4
    minichromosomes, 71–80, 142
    in *Xenopus* oocyte, DNA from,
      fate, 73–4

T
T antigen, SV40, 95, 103–4
T7 DNA, transcription, 164
T7 RNA polymerase, 165
Tails, histone, 12
    deletions in, effects, 149
    DNA organization in
      nucleosomes and, 25–6
    post-translational modification,
      54–5, 56–7
    SPKK (S/T P-X-K/R) motifs in, 52,
      58
TATA box, 106–7, 143, 148–9
Telomere DNA, proteins associated
    with, 47–8
Testis, histone variants, 62
Tetrahymena, mitosis and
    chromosome compaction in,
    uncoupling, 58–9
TFIIB, 107
TFIID, 106–7, 130, 143, 144
TFIIF, 110
TFIIIA, 108–10, 116–17, 132–3
TFIIIB, 108–10
TFIIIC, 108–10, 116–17
Thymidine kinase promoter, HSV,
    105–6
Thymine, *see also* Oligo(dA.dT)
    tracts
  base pairing, 5, 7
  structure, 5

TIF-1, 110
Topoisomer(s), of DNA, 9–10
Topoisomerase I, activity/role, 41,
    44
Topoisomerase II, activity/role, 41,
    44–5, 45–6, 98
TP 2, chromatin remodelling and,
    61–2
Trans-acting factors, 3, 123–50, 169,
    see also specific factors and
    Non-histone/DNA-binding
    proteins
    chromosomal/chromatin structure
      and, 140–50
    local organization of, 147–50
    specific, 126–40
      non-specific chromatin and,
      126–31
      specific chromatin and, 131–40
Transcription, 69, 106–66
    chromatin integrity and, 160–6
      in vitro, 164–6
      in vivo, 160–4
    chromatin structure incompatible
      with, incorrectly assembled,
      76
    of chromatin/DNA (in general),
      106–66
      basic machinery, 106–23
      H1–repressed, in erythrocytes,
      65
    histone acetylation and,
      correlation, 56
    developmental variations, *see*
      Development
    DNA replication and its
      consequences for, 118, 119,
      151–9
    histone genes, 51–4
    regulation, 116–23
    of somatic nuclei in *Xenopus*
      oocytes, 70
Transcription complexes, 111, 112–
    20, 153, 155–9
    maintenance through DNA
      replication, 118, 119, 153,
      154, 155–9
    stable, 111–16
Transcription factors, 104, 106–10,
    116–17, 130, 132–3, 144–5,
    see also specific factors

non-histone structural proteins
and, homology, 66–7
Transfer RNA, transcription, 107,
109
Transition proteins, chromatin
remodelling and the, 61–2
Translation, histone modification
following, 53, 54–61, 86–7
Translational position of
nucleosomes, 29
Transplantation, nucleus, 69–71
TRP1ARS1, 82–4, 139–4
Tyrosine aminotransferase gene
enhancer, 137

U
UBF, 61–7, 110
Ubiquitination of histones, 57
Ubx gene, 146–7
UNF region, 82, 83
Upstream binding factor, 110

V
Viral systems, replication and
transcription in, 103–4, 158,
see also specific viruses

X
X chromosome, inactive,
heterochromatic, 65
Xenopus spp.
5S RNA gene
chromatin assembly and the,
74–5, 154
nucleosome position and the,
29–30
transcription, 29–30, 116–17,

132–3, 154, 157
DNA replication and
transcription during
development, 158
eggs/oocytes
chromatin assembly in, 73–8,
154–5, *see also* Eggs; Oocytes
DNA synthesis in, 154–5
nuclear assembly in extracts of,
96–9
nuclear transplantation
experiments, 69–71
nucleosome assembly in
extracts of, 90–5
H1 variants, 52–4
organization into chromatin of
genes from, 74–6
X-ray diffraction studies of DNA
structure, 6, 8

Y
Y-box, cold-shock domain of, 121
Yeast
DNaseI-sensitive/hypersensitive
sites, 148, 162
heat shock proteins, 143, 146, 162
mating type, 84, 148–9
minichromosomes, 82–5, 139–40,
148
transcription, 133–4, 143, 146,
148–9, 162

Z
Z-DNA, 8
Zigzag arrays of nucleosomes, 36
Zigzag structure of chromatin
fibers, 32–3, 34
Zinc finger proteins, 64, 121